Essential Maple 7

Springer

New York
Berlin
Heidelberg
Barcelona
Hong Kong
London
Milan
Paris
Singapore
Tokyo

Robert M. Corless

Essential Maple 7

An Introduction for Scientific Programmers

With 76 Illustrations

 Springer

Robert M. Corless
Department of Applied Mathematics
University of Western Ontario
London, Ontario N6A 5B7
Canada
Rob.Corless@uwo.ca

Mathematics Subject Classification (2000): 68-04

Library of Congress Cataloging-in-Publication Data
Corless, Robert M.
    Essential Maple 7 : an introduction for scientific programmers / Robert M. Corless.
        p.   cm.
    Includes bibliographical references and index.
    ISBN 0-387-95352-3 (alk. paper)
    1. Maple (Computer file)   2. Mathematics—Data processing.   I. Title.
QA76.95 .C678 2001
510′.2855369—dc21   2001048432

Printed on acid-free paper.

Maple is a registered trademark of Waterloo Maple Inc.

Production managed by Michael Koy; manufacturing supervised by Jacqui Ashri.
Typeset by Integre Technical Publishing Company, Inc., Albuquerque, NM.
Printed and bound by Edwards Brothers, Inc., Ann Arbor, MI.
Printed in the United States of America.

9 8 7 6 5 4 3 2 1

ISBN 0-387-95352-3   SPIN 10850813

Springer-Verlag   New York Berlin Heidelberg
A member of BertelsmannSpringer Science+Business Media GmbH

For my parents:

John D. Corless and Marion L. Corless
and
M. Aly Hassan and Galima Hassan

What's in This Book

This book contains an accelerated introduction to Maple, a computer algebra language. It is intended for scientific programmers who have experience with other computer languages such as C, FORTRAN, or Pascal. If you want a longer and more detailed description of how to program in Maple, see [44].

The mathematical prerequisites are calculus, linear algebra, and some differential equations. A course in numerical analysis will also help. Any extra mathematics needed will be developed in the book.

This book was originally prepared using an earlier version of Maple, but has been revised for Maple Release 7, with an eye towards changes for the next release after that. Maple continues to be an evolving system. New features will be described in the documentation for updates (`?updates` in Maple), and any necessary updates of the text of this book will be made available over the Web. See my web page http:// www.apmaths.uwo.ca/ ~rcorless for a pointer.

Indeed, one reason that there was so much time between the first and second editions of this book is precisely that Maple has been evolving so rapidly in the last few years, too rapidly for me to revise this book (much less complete my others) while coping with my other duties. Maple is now a substantially better product than it was, with important improvements to the programming language itself (particularly, nested lexical scopes and modules) and to the library of "black boxes" (particularly `LinearAlgebra`). This book takes complete account of these improvements: All the programs and examples and exercises in this book have been revised, many quite substantially. The former Chapter 4, which was a subject-oriented keyword summary of Maple, has been supplanted completely by the on-line help system, and therefore cut from the book. In spite of cutting that chapter, the total number of pages in the book has increased for the second edition, because much new material has been added, including an appendix on complex variables in a computer algebra context.

In spite of the additions, this book does not provide complete coverage of Maple. For example, I don't talk about so-called "smart" plots, or about the facilities for exact solution of partial differential equations. Without doubt, some readers would find it useful for me to write about some of these omitted topics. On the other hand, also without doubt, I have included some topics that are only needed by some readers, not all. The topic selection is a compromise, and I hope that you don't mind those selections I have made that don't fit your needs. Please send me your suggestions for topics to include in the next edition, or in the electronic updates.

This book does not require any particular hardware. The systems I have used in developing the book are machines running Windows 98 and Windows NT, Linux machines, and X-windows systems. There should be no adjustments necessary for readers equipped with Macintoshes or other hardware.

How to Read This Book

The suggested way to read this book is to read Sections 1.1–1.3 at a sitting, while you have Maple running in front of you so you can try things out. Read the rest of the book at your leisure, and in any order you like.

There are many small programs scattered throughout this book, and I hope that you may find them useful in themselves, and as guides for writing your own.

The exercises are intended to give you practice in what has just been shown and to develop the ideas further. They vary in difficulty from trivial to quite difficult. They have been used as assignments in an introductory graduate course in applied computer algebra here at the University of Western Ontario. It is not necessary to do them to benefit from this book, but it is probably more fun than just reading. I plan, with the help of some of my students, to provide a solutions manual; see my web page http://www.apmaths.uwo.ca/~rcorless for details.

Acknowledgements

The most significant help I received for this book was from my wife, Sumaya. By hard work in a wide variety of capacities she has made writing this book both possible and very pleasant. She deserves much more credit than she gets from this one little paragraph. This remains true for the second edition.

In the acknowledgments for the first edition of this book, I wrote:

> My daughter Shamila, on the other hand, hasn't really been helpful at all—but she always *wants* to help, and somehow that's just as good, from someone who's five years old.

Well, she's now a little older, and she still wants to help; now, she does.

My parents (both sets), to whom this book is dedicated, provided me with an invaluable foundation from which to work.

For technical help, thanks go to Keith Geddes for getting me interested in computer algebra with the first Maple course offered at the University of Waterloo. The course began with ALTRAN and finished with Maple (this was back before version numbers). I have since used computer algebra in nearly all my work, both research and teaching. The other major influence on my computer algebra career is David Jeffrey, who taught me what it means to do research in applied mathematics and has continued as a good friend and collaborator. The members of the watmaple mailing group, past and present, from Gaston Gonnet and Michael Monagan through to the most recent student research assistant, have participated in many extremely interesting discussions and have taught me a lot about Maple. George Labahn gave some interesting examples of Maple plots and consulted on several aspects. George Corliss, Dave Hare, Henning Rasmussen, and Kelly Roach provided particularly detailed criticisms of early drafts of the first edition of this book. Bill Bauldry and an anonymous reviewer also provided helpful remarks. Niklaus Mannhart helped with some final LaTeX work, as well as reading over the manuscript.

Portions of the first edition of this book were finalized while I was on sabbatical at T. J. Watson Research Center in Yorktown Heights, New York. Stephen Watt, Dick Jenks, and Tim Daly were generous with their time and energy, even while on a tight schedule.

Thanks also go to Darren Redfern for excellence as an "author's editor"; to Stan Devitt for developing the processing tools used to efficiently include Maple input and output in this LaTeX document; to my students Mohammed O. Ahmed, Anne-Marie E. Allison, Tianhong Chen, and David W. Linder for being "lab animals" in testing this book; and similarly to all my applied computer algebra students, past and present, for helping me to refine my ideas and presentations.

I never thanked Betty Sheehan of Springer for all her help with the first edition; I do so now.

Acknowledgements for help with the second edition.   My Ph.D. students Xiaofang Xie and Mhenni Benghorbal read the first edition carefully, and suggested many places for improvement and alignment with Maple 7. They also did all the exercises. My former Ph.D. student Xianping Liu also found some bugs in a draft of this book.

My colleague Jacques Carette at Waterloo Maple Inc. has provided detailed criticism of this edition of the book, and invaluable advice on Maple programming and capabilities. Without him, I would not know about many of the interesting programming features of Maple. Other people at WMI who have provided helpful remarks include Douglas Harder and Paulina Chin. Karen Ranger turned all the worksheets from this book into "Power Tools" (see p. 16).

Cleve Moler took some time to glance at the section in this book on the MATLAB link. His remarks were very helpful, though I haven't yet tried out the multidimensional FFT as he suggested. Thanks also go to Arthur Norman, for his "cheerful sniping."

Achi Dosanjh and the staff at Springer have been very helpful with the production of this book (which because of its tight schedule, once the Maple 7 beta got into my hands, has put a strain on everyone). I am very grateful for the improvements to this book that they have made possible, and particularly for the astute comments of David Kramer, who provided copy-editing. Any errors that remain are of course my fault.

Finally, thanks go to Mark Giesbrecht, David Jeffrey, Greg Reid, and Stephen Watt, my colleagues at the UWO branch of the Ontario Research Centre for Computer Algebra (ORCCA) for the past two years. They provide, first of all, friendship, and, second, an extremely satisfying research and teaching environment to work in. They make ORCCA at UWO into the best possible place for me to do computer algebra.

London, Canada   Robert M. Corless
http://www.orcca.on.ca

Contents

List of Figures

1

Basics

> ' . . . But the mad will ne'er content, till he shall have patterned
> out to his own most mathematical likings the unpeerable inventions
> of God . . . '
> —E. R. Eddison, *A Fish Dinner in Memison*, Chapter IX.

This chapter shows how to get started in Maple,[1] gives some sample sessions, discusses some common difficulties and errors, and lays a firm groundwork for more advanced use of Maple. Most important are Sections 1.1 to 1.3; the others can be read after you skim the rest of the book.

This book has been tested on Windows and Unix systems; some differences may appear from session to session, and some details of the commands are different for Macintosh systems.

1.1 Getting Started

To start Maple on windowing systems, double-click on the Maple icon, or type `xmaple` at a Unix command prompt. This will bring up a graphical user interface window, called a *worksheet*, that gives access to all of Maple's commands. Consult your local wizard if this doesn't work. To get help once you have started Maple, type ? after the Maple prompt, which is usually an angle bracket (>), and hit RETURN or ENTER. For example, what appears in Figure 1.1 is what you get if you type `?index` after the prompt. To get help on a particular Maple topic, type `?keyword` where `keyword` is the Maple word for what you want help on.

---

[1] Maple is a registered trademark of Waterloo Maple Incorporated.

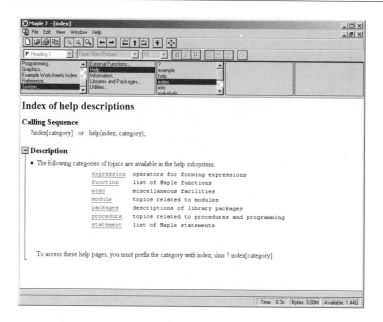

Figure 1.1: The result of the command `?index`

If you don't know the exact word, simply guess a few alternatives: Maple will try to help you locate what you want. As long as you're not using the command-line version of Maple, which doesn't have a fancy graphical user interface, you can use the *help browser* to try to search by category, or even by "full text search."

To use the help browser, open Maple and then click on the `Help` menu item (on the far right of the top line). Maple's "help entries" contain worked examples and links to other pages, as well as references to mathematical material. Navigation in the help browser is rather similar to navigation in an Internet browser, but it isn't exactly the same (because, naturally enough, it was done first in Maple).

Exercises

1. Work through the "New User's Tour" in the help browser. This will take you about two hours if you do everything.

2. Find out how to use the linear algebra package by starting Maple and issuing the commands `?LinearAlgebra` and `?with`. Explore at least one routine (e.g., `JordanForm`).

3. If you have MATLAB[2] on your machine, find out how to use the MATLAB link from Maple by issuing the command ?Matlab. Explore at least one routine (e.g., evalM). You may have to configure your system, as advised in the help file. **Aside:** I highly recommend the beautiful book [30] to all MATLAB users.

1.1.1 Basic Command Syntax

Note that Maple is *case-sensitive*, so series is different from SERIES is different from Series. The Maple command is series.

```
> series( sin(exp(x)-1), x );
```

$$x + \frac{1}{2} x^2 - \frac{5}{24} x^4 - \frac{23}{120} x^5 + O(x^6)$$

That is one way to compute a series in Maple. But if instead we use some uppercase letters, Maple thinks we're talking about some other function that it may learn about later:

```
> Series( sin(x), x );
```

$$\text{Series}(\sin(x), \, x)$$

As you see, Maple echoes syntactically legal input that it doesn't understand. This behaviour is fundamental to Maple's ability as a symbolic processor, but in this case it may not be what is wanted. In particular, if you have the "caps lock" key on initially, you may get something like the following.

```
> SERIES( sin(x), x );
```

$$\text{SERIES}(\sin(x), \, x)$$

Maple statements end with a semicolon (;) or colon (:). Statements ending in a colon (:) perform computations, but the results are not printed. This is used to suppress the printing of voluminous intermediate results. For example,

```
> expand( (x+y)^3 );
```

$$x^3 + 3 x^2 y + 3 x y^2 + y^3$$

displays its results, as expected. On the other hand, suppose we wish to compute the coefficient of x^{48} in $(x + 3)^{100}$. Then the intermediate result below is not of any real interest, and since it occupies 262 lines on my screen it is a good idea not to print it.

```
> DoNotLookAt := expand( (x+3)^100 ):
> coeff( DoNotLookAt, x, 48 );
```

$$6022152095047244121129504670562748296126762243770345100$$

---

[2]MATLAB is a registered trademark of The MathWorks, Inc.

Instead of referring to the variable DoNotLookAt in the second command, I could have used the % variable, which (in versions of Maple later than Release 4) refers to the previous result: coeff(%, x, 48);.

The use of the percent variable replaces the (Release 4 and earlier) use of ditto (") for the same concept.

In this edition of the book I will use the % variable infrequently, not because of incompatibility, but rather because I now believe that exposition is usually clearer without it. It is generally true both that worksheets are easier to read, and that Maple programs are more efficient, if they are written without using %. On the other hand, % is concise and useful for "on the fly" calculations, especially for results that you will throw away afterwards. In such cases, I use % in this book.

A common mistake: If you forget to type the statement terminator (a colon or semicolon), you will get a warning message:

```
>   DoNotLookAt := expand( (x+3)^100 )

Warning, premature end of input
```

Do not retype the line; simply enter a colon or semicolon and hit return. In a worksheet you may go back to the end of the line in question and enter it there. If instead you retype the line (so that you see two copies of your command on the screen) you will (probably) introduce a syntax error, because Maple will try to interpret what you have typed twice as a single, multiline, Maple statement. Other common mistakes are covered in Section 1.2.2.

1.1.2   Use of Context-Sensitive Menus to Execute Maple Commands

Many Maple commands can be executed without typing them. Simply right-click on the Maple result that you wish to manipulate, and a menu of possible actions will appear. By selecting from one of the actions, the appropriate Maple command will be typed for you and executed. See Figure 1.2.

1.1.3   Sample Maple Sessions

Three short Maple sessions follow. You should start Maple up on your system, and type in the following commands.

First Sample Session: Maple as Calculator

This session shows how to use Maple to solve some problems in algebra, linear algebra, and calculus. We begin by factoring a polynomial, but before we begin we restart the Maple session:

```
>   restart:
```

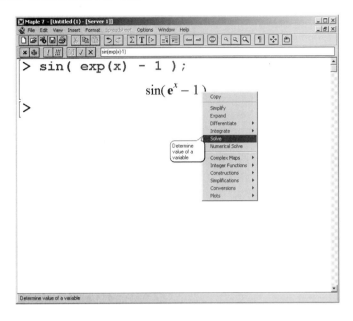

Figure 1.2: Context-sensitive menus are available by right-clicking

The purpose of `restart` is to give us a fresh Maple session, and putting it at the start of a worksheet helps when we execute the worksheet again.

Now we may begin our actual session. It is possible that the ordering of the factors output below may be different in your session.

> `FactoredForm := factor( t^12 - 1 );`

FactoredForm :=
$$(t-1)\,(t^2+t+1)\,(t+1)\,(1-t+t^2)\,(t^2+1)\,(t^4-t^2+1)$$

Is that factoring correct? We can see by inspection that the roots ± 1 are included in those factors, as are the roots $\pm i$ (where i is the square root of -1). So we tend to believe that Maple got that factoring right, and of course we can ask Maple to expand that factoring out to get back $t^{12} - 1$.

> `expand( FactoredForm );`
$$t^{12} - 1$$

Now let us do some simple computations from linear algebra, using the
LinearAlgebra module,[3] which was introduced in Release 6 and ultimately is
to replace the linalg package. This book will not use the old linalg package
unless it is necessary.

> with(LinearAlgebra):

The call with(LinearAlgebra) enables simple access to Maple's linear algebra
package.

> A := Matrix([[4,5], [5,6]]);

$$A := \begin{bmatrix} 4 & 5 \\ 5 & 6 \end{bmatrix}$$

In Maple, one uses a Matrix to represent a mathematical matrix, and Vector
to represent a vector. This is different from an Array or the older matrix and
array. See ?Matrix for details. I will always use capitals for matrices and vectors in my descriptions in this book to emphasize that the new Maple representations are to be preferred. Matrix multiplication (noncommutative multiplication)
is written in Maple with the . operator.

> A . A ;

$$\begin{bmatrix} 41 & 50 \\ 50 & 61 \end{bmatrix}$$

Now compute the characteristic polynomial of the matrix A.

> CharacteristicPolynomial(A, lambda);

$$-1 - 10\lambda + \lambda^2$$

The trace of the matrix is $6 + 4 = 10$, which should be the negative of the linear
coefficient. It is. The determinant of the matrix is $4 \cdot 6 - 5^2 = -1$, which should
be the constant coefficient. Again, it is. As before, we conclude that Maple got it
right.[4]

Now let us do some calculus (again, using restart to give us a fresh session):

> restart:

> Int(1/(t^6-1), t) = int(1/(t^6-1), t);

The output from this command appears below. Let us first examine the input. Note
that the left-hand side of the input is the same as the right-hand side, except that
the left-hand side has a capitalized Int (which makes the command *inert*. We
will discuss inert functions in Section 1.6.2). Here it is used to produce a sensible

---

[3]A Maple programming language module has no relation to a module from algebra. It is one programming construct that
can be used to implement a package of routines with a common theme. See Chapter 3 for a discussion of Maple modules.

[4]It is always good policy to make simple checks on results. Because mathematics is so rich, there are usually more options
to do so with the results of computer algebra systems than with other computer programs.

equation, with an integral on the left-hand side of the output below, put equal to an expression on the right-hand side.

$$\int \frac{1}{t^6 - 1} \, dt = \frac{1}{6} \ln (t - 1) - \frac{1}{6} \ln (t + 1)$$
$$+ \frac{1}{12} \ln \left(t^2 - t + 1\right) - \frac{1}{12} \ln \left(t^2 + t + 1\right)$$
$$- \frac{\sqrt{3}}{6} \arctan \left(\frac{\sqrt{3}}{3} (2t + 1)\right) - \frac{\sqrt{3}}{6} \arctan \left(\frac{\sqrt{3}}{3} (2t - 1)\right)$$

Note that Maple did not add an arbitrary constant to its answer. This is supposed to be understood, and if you wish to have the constant there explicitly, you must put it in yourself by adding it on, as in (for example)

```
>  int( 1/(t^6-1), t ) + C;
```

which would add the constant C to the computed answer.

The computed answer to $\int 1/\left(t^6 - 1\right)$ looks formidable. If we wish to check that answer independently from Maple, we would most likely prefer to do it numerically. However, the code in Maple for differentiation is independent of the code for integration, so if we ask Maple to differentiate both sides of the above equation we will get a useful confirmation.

```
>  diff( %, t );
```

$$\frac{1}{t^6 - 1} = \frac{1}{6} \frac{1}{t - 1} - \frac{1}{6} \frac{1}{t + 1}$$
$$+ \frac{1}{12} \frac{(2t - 1)}{t^2 - t + 1} - \frac{1}{12} \frac{2t + 1}{t^2 + t + 1}$$
$$- \frac{1}{3} \frac{1}{1 + \frac{1}{3} (2t + 1)^2} - \frac{1}{3} \frac{1}{1 + \frac{1}{3} (2t - 1)^2}$$

Note that the symbol % refers to the last result, as discussed before. The last result was an *equation*, with two sides. Both sides were differentiated: Now we have only to simplify the results. It turns out, after some experimentation, that the command that simplifies things most efficiently is normal with the expanded option.[5]

```
>  normal( %, expanded );
```

$$\frac{1}{t^6 - 1} = \frac{1}{t^6 - 1}$$

---

[5]Try it without the option. You will see that the resulting equation is simpler than before, but not as simple as given here; you will also see that this option is a fairly natural thing to try in this case.

So it appears that Maple found a correct antiderivative for $1/\left(t^6 - 1\right)$. Now let us solve the logistic differential equation.

```
>   restart:
```

```
>   Logistic := diff( x(t), t ) = x(t)*(1-x(t));
```

$$Logistic := \tfrac{\partial}{\partial t}\, x(t) = x(t)\,(1 - x(t))$$

Let us use the initial condition $x(0) = \alpha$.

```
>   initialCond := x(0) = alpha;
```

$$initialCond := x(0) = \alpha$$

Now solve the differential equation with this initial condition for the unknown $x(t)$.

```
>   ans := dsolve( {Logistic,initialCond}, x(t) );
```

$$ans := x(t) = \frac{1}{1 - \dfrac{e^{(-t)}\,(-1 + \alpha)}{\alpha}}$$

Many people are more skilled than Maple is at writing things concisely. Such a person might prefer to write that solution as

$$x(t) = \frac{\alpha}{\alpha + (1 - \alpha)e^{-t}}\; .$$

In addition to being neater, that formulation has the advantage of being correct when $\alpha = 0$, whereas the result returned from Maple has a spurious problem at $\alpha = 0$. We discuss the difficulties of automatic simplification, and of specialization and continuity, later in this book. We check the result both by substituting it back into the differential equation and by verifying the initial condition.

```
>   check := eval( Logistic, ans );
```

$$check := -\frac{e^{(-t)}\,(-1 + \alpha)}{\left(1 - \dfrac{e^{(-t)}\,(-1 + \alpha)}{\alpha}\right)^{2}\,\alpha} = \frac{1 - \dfrac{1}{1 - \dfrac{e^{(-t)}\,(-1 + \alpha)}{\alpha}}}{1 - \dfrac{e^{(-t)}\,(-1 + \alpha)}{\alpha}}$$

The functions lhs and rhs stand for "left-hand side" and "right-hand side," respectively.

```
>   normal( lhs(check)-rhs(check) );
```

$$0$$

Since the left- and right-hand sides are equal, $x(t)$ is indeed a solution. Let's do that again, a little more slowly, using subs:

```
> check := subs( ans, Logistic );
```

$$check := \frac{\partial}{\partial t} \frac{1}{1 - \dfrac{e^{(-t)}(-1+\alpha)}{\alpha}} = \frac{1 - \dfrac{1}{1 - \dfrac{e^{(-t)}(-1+\alpha)}{\alpha}}}{1 - \dfrac{e^{(-t)}(-1+\alpha)}{\alpha}}$$

The subs command substitutes but does not evaluate. Thus we can see an intermediate step that we missed before.

Now we compare the left-hand side with the right-hand side again, but this time we use *arithmetic of operators* (see Chapter 3):

```
> normal( (lhs-rhs)(check) );
```

$$0$$

That construct, applying the operator lhs-rhs to the equation, is preferred because we need to type the name of the equation only once. Another reason to use this construct is that we could map it, should we choose, onto a list or set of equations.

Now check the initial conditions.

```
> checkIC := normal( subs(t=0,ans), expanded );
```

$$checkIC := x(0) = \alpha$$

Therefore, the initial condition is satisfied.

This process can be carried out automatically using the odetest command. See ?odetest.

```
> odetest( ans, Logistic );
```

$$0$$

Terminating a Maple Session

You quit a Maple session by issuing the command quit, done, or stop, which leaves Maple running but closes your current worksheet, or by choosing Exit from the menu (which shuts Maple down completely) if you are using a menu-driven system. These statements can be terminated by a carriage return or enter; no colon or semicolon is necessary. Now quit the Maple sample session.

```
> quit
```

This ends the first sample session. You can get help on the meaning of each of the commands used above by the ? command, as noted previously.

Exercises

1. Factor $t^{24} - 1$. Check your answer.

2. Factor $t^6 - 1$ down to linear factors. Hint: See ?factor, and you need the extension $K = (-3)^{1/2}$, though $\sqrt{-3}$ does not work because it gets imme-

diately converted to the product $i\sqrt{3}$; you can use the set of two extensions $\{i, \sqrt{-3}\}$ though. Again (and always) check your answer.

3. Find the inverse of the matrix A from this sample session, using Maple (see `?MatrixInverse`).

4. Maple also implements *elementwise* operations on collections of data, by using an `Array` instead of a `Matrix`. Convert the `Matrix` A of this section to an `Array` by issuing the command

   ```
   >  B := Array( A );
   ```

   or the more efficient, because "in-place," operation

   ```
   >  rtable_options( A, subtype=Array ):
   ```

   Then, compute B^{-1}, $B * B$, and B . B and compare the results to operations on Matrices. The elementwise product is known (following von Neumann) as the Hadamard product. In MATLAB one uses the . * operator. Componentwise operations can also be done efficiently using the zip and Zip functions: Zip('/',A,B) divides every element of the (Array or Matrix or Vector) A by the corresponding element of the (Array or Matrix or Vector) B.

5. Find $\int e^{-t} \sin t \, dt$.

6. Find the fifth derivative of $e^{\theta} \arcsin\theta$ (the inverse sine function is called `arcsin` in Maple).

7. Solve $x^2 y'' + x y' + y = 3x^3$ for $y(x)$.

Second Session: 'Hello, World'

For the second sample session we write the obligatory "Hello, World" program (or a slight variant of it). Start Maple again, and enter the following one-line program.

```
>  restart:
```

```
>  hi := proc() "Hello, Worf." end proc;
```

The quotes are string quotes ("), not left quotes (') or right quotes (').

$$hi := \mathbf{proc}() \text{``Hello, Worf.''} \ \mathbf{end \ proc}$$

```
>  hi( );
```

                                "Hello, Worf."

Maple procedures can accept any number of arguments, although here no arguments are actually used in the procedure.

```
>  hi( Anything );
```

                                "Hello, Worf."

Now quit Maple again.

```
>  quit
```

A Maple procedure body is begun with a `proc()` keyword and ended with the end keyword. In Release 6 or later you may say end `proc` to help you know just what it is you are ending. A procedure takes arguments, which may or may not be indicated in the `proc()` keyword. To give a name to a procedure, you assign the procedure body to the name. In the above example, the name of the procedure was `hi`.

Exercises

1. Compare the results of the following Maple input statements (`hi` is defined as above).

```
>  hi;
>  hi( Throgmorton );
>  hi( Ginger );
```

2. What will the following program do? Entering this short a program in a worksheet is acceptable (longer programs should be done with an editor and read into Maple, by the `read` command). You can insert a line break into a program in a worksheet by holding down the SHIFT key while you hit RETURN.

```
>  hi := proc( x :: string )
>      if x="Maple" then
>          "Hello, yourself.";
>      else
>          "I beg your pardon?";
>      end if;
>  end proc:
```

Third Session: Heat Conduction by Fourier Series

For our final sample session we attempt something a little more ambitious, namely, the solution of the heat equation $u_t = u_{xx}$ on $0 \leq x \leq 1$, for $t > 0$, with boundary conditions $u(0, t) = u(1, t) = 0$, and initial condition $u(x, 0) = f(x)$, say $f(x) = x^2(1 - x^2)$. This will give us the nondimensional temperature u for all later times t in a rod of length 1 whose initial temperature is distributed as $x^2(1 - x^2)$.

We use the standard theory of Fourier series to solve this problem (see, for example, [4]) and use Maple as a worksheet for the calculations. This gives a quick overview of integration, some algebraic manipulation facilities, and some elementary plotting features. However, it does use a few advanced features of Maple. These may appear to be somewhat mysterious at this stage. I urge you to follow along through the sample session as far as you can, and skim until the end of the session if you get into trouble and can't get out; you can always return to this example later. As this session proceeds, issue help commands as needed (for example, ?int).

According to any reference book on Fourier series, for example the excellent [40], the solution that we are looking for can be written in the form

$$u(x, t) = \sum_{k=1}^{\infty} c_k e^{-k^2 \pi^2 t} \sin(k\pi x),$$

and we will use Maple below to calculate the constants c_k. It is known that

$$c_k = \frac{\int_0^1 f(x) \sin(k\pi x)\, dx}{\int_0^1 \sin^2(k\pi x)\, dx},$$

so we begin with the evaluation of these integrals.

```
>   restart:
```

Note the use of the *range* to represent the interval $0 \le x \le 1$.

```
>   I1 := int( sin(k*Pi*x)^2, x=0..1 );
```

$$I1 := \frac{1}{2} \frac{-\cos(k\pi)\sin(k\pi) + k\pi}{k\pi}$$

It seems obvious to us that the above expression can be simplified, but we must remember that Maple does not know that k is an integer. The simplest way to help Maple out is to use our own knowledge. Later, we will see a way to get Maple to assume properties of variables, such as that $k > 0$ is an integer.

```
>   I1 := eval( I1, {sin(k*Pi)=0,cos(k*Pi)=(-1)^k} );
```

$$I1 := \frac{1}{2}$$

Of course, the evaluation of $\cos(k\pi)$ was wasted, but I included it to show that one can perform multiple evaluations or substitutions at once. This gives us one of the integrals in the definition of c_k. Now for the other.

```
>   I2 := int( x^2*(1-x^2)*sin(k*Pi*x), x=0..1);
```

$$I2 := -2(-12\cos(k\pi) + k^3\pi^3\sin(k\pi) + 5k^2\pi^2\cos(k\pi)$$
$$- 12k\pi\sin(k\pi) + k^2\pi^2 + 12)\Big/(k^5\pi^5)$$

```
>   I2 := eval( I2, {sin(k*Pi)=0, cos(k*Pi)=(-1)^k} );
```

$$I2 := -2\frac{-12(-1)^k + 12 + 5k^2\pi^2(-1)^k + k^2\pi^2}{k^5\pi^5}$$

> I2 := collect(I2, k, factor);

$$I2 := -2\,\frac{5\,(-1)^k + 1}{\pi^3\,k^3} + \frac{24\,((-1)^k - 1)}{\pi^5\,k^5}$$

We will take up the difficult question of how to get Maple to simplify its results in Chapter 2. Continuing with this example, now we know that c_k is the above value divided by $\frac{1}{2}$. The following is a convenient way to express this as a *functional operator*. See Section 3.2 for more details on operators.

> c := unapply(2*I2, k);

$$c := k \to -4\,\frac{5\,(-1)^k + 1}{\pi^3\,k^3} + \frac{48\,((-1)^k - 1)}{\pi^5\,k^5}$$

The name unapply is not evocative of the procedure's purpose until you think of *applying* an operator, say $g : k \mapsto k^3 + 5$, to an argument, say m, to get the *expression $m^3 + 5$*. The opposite of this process, namely, converting an expression to an operator, is then reasonably thought of as "unapplication."

Returning to our example, we can express the sum to n terms of the Fourier series for the initial condition $f(x)$ as follows. We use the add command, and not the sum command, because the sum command is overly powerful, intended to do "symbolic summation," whereas the add command just adds terms. Note the use of the range to represent the set of integers $1, 2, \ldots, n$. Compare this with the use of ranges earlier to represent a real interval.

> fn := n -> add(c(k)*sin(k*Pi*x), k=1..n);

$$fn := n \to \text{add}(c(k)\sin(k\,\pi\,x),\; k = 1..n)$$

Let us take five terms in the series and investigate the error.

> f5 := fn(5);

$$f5 := \left(16\,\frac{1}{\pi^3} - \frac{96}{\pi^5}\right)\sin(\pi\,x) - \frac{3\sin(2\,\pi\,x)}{\pi^3}$$
$$+ \left(\frac{16}{27}\,\frac{1}{\pi^3} - \frac{32}{81}\,\frac{1}{\pi^5}\right)\sin(3\,\pi\,x) - \frac{3}{8}\,\frac{\sin(4\,\pi\,x)}{\pi^3}$$
$$+ \left(\frac{16}{125}\,\frac{1}{\pi^3} - \frac{96}{3125}\,\frac{1}{\pi^5}\right)\sin(5\,\pi\,x)$$

> plot(f5 - x^2*(1-x^2), x=0..1);

See Figure 1.3. The maximum magnitude of the error is less than 0.01, with only five terms in the series. The maximum value of the function occurs when $x^2 = 1 - x^2$ (by the arithmetic–geometric mean inequality [46], or by elementary calculus) or $x^2 = \frac{1}{2}$ so $f(x) = \frac{1}{4}$ at this point. Thus the relative error is less than 5%. Let us see what happens if we take 10 terms. Here we don't want to *look* at the ten terms in the series, just compute them.

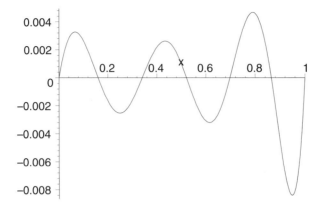

Figure 1.3: Graph of the error in the five-term solution to the heat equation

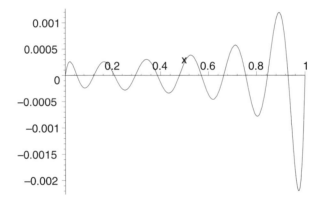

Figure 1.4: Graph of the error in the ten-term solution to the heat equation

```
>   f10 := fn( 10 ):
>   plot( f10 - x^2*(1-x^2), x=0..1 );
```

See Figure 1.4. The maximum magnitude of the error now is roughly 0.002, or less than 1%.

Now let us consider the solution u. Does it satisfy the differential equation? The Fourier series was constructed so that each term *should* satisfy the differential equation, so the only error we are expecting to commit is in the representation of the initial condition. We check that our solution satisfies the differential equation, to guard against blunders,[6] as opposed to approximation errors.

```
>   un := n -> add( c(k)*exp(-k^2*Pi^2*t)*sin(k*Pi*x),
>       k=1..n );
```

$$un := n \to \mathrm{add}(c(k)\,e^{(-k^2\pi^2 t)}\sin(k\,\pi\,x),\ k = 1..n)$$

```
>   u5 := un( 5 );
```

$$u5 := \left(16\,\frac{1}{\pi^3} - \frac{96}{\pi^5}\right) e^{(-\pi^2 t)} \sin(\pi\,x) - \frac{3\,e^{(-4\pi^2 t)}\sin(2\,\pi\,x)}{\pi^3}$$

$$+ \left(\frac{16}{27}\,\frac{1}{\pi^3} - \frac{32}{81}\,\frac{1}{\pi^5}\right) e^{(-9\pi^2 t)} \sin(3\,\pi\,x) - \frac{3}{8}\,\frac{e^{(-16\pi^2 t)}\sin(4\,\pi\,x)}{\pi^3}$$

$$+ \left(\frac{16}{125}\,\frac{1}{\pi^3} - \frac{96}{3125}\,\frac{1}{\pi^5}\right) e^{(-25\pi^2 t)} \sin(5\,\pi\,x)$$

```
>   diff( u5, t ) - diff( u5, x, x );
```
$$0$$

Therefore, the five-term approximation satisfies the differential equation exactly.

```
>   u10 := un( 10 ):
>   diff( u10, t ) - diff( u10, x, x );
```
$$0$$

Likewise, the ten-term approximation satisfies the differential equation exactly.

We should also check that the boundary conditions $u(0, t) = u(1, t) = 0$ are satisfied, but this is obvious from the fact that each term is multiplied by $\sin(k\pi x)$ for some integer k.

From the previous calculation of f_5 and f_{10} we know that the initial condition is not satisfied exactly, but only approximately.

Now let us draw a contour plot of this function u of x and t. After some experimentation, we find that the following scale gives useful information.[7]

```
>   plots[contourplot]( u10, x=0..1, t=0..0.21,
>       grid=[30,30], colour=black );
```

That plot, shown in Figure 1.5, shows *isotherms*, or lines of equal temperature. We see that for small times and x less than about 0.4, the temperature rises initially and then falls. This makes sense, because the initially peaked temperature distribution (which is not symmetric, and has its maximum at $x = 1/\sqrt{2}$) is smoothed

---

[6]A blunder is an accidental mistake, as opposed to an approximation error, which is merely a compromise. The word "blunder" survives mostly in the chess community nowadays, but famous examples, such as the lines "Not tho' the soldier knew/Some one had blunder'd" in Tennyson's "Charge of the Light Brigade," still remain (thanks go to David Kramer for pointing that one out).

[7]The earlier plots had their y-scales chosen automatically; here, we have two variables x and t to scale, and the choice of the t-scale is not obvious.

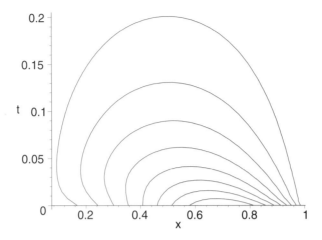

Figure 1.5: Lines of equal temperature in the *x*-*t* plane

out as time progresses and the initially colder parts of the rod are warmed by the hotter adjacent region.

Now suppose we wished to repeat this process for several different initial functions $f(x)$. It would make sense to write a Maple program to automate those steps. This is probably the main strength of Maple, or indeed of any computer algebra language: It is a high-level programming language, which you can customize to your own needs. One version of such a program can be found in Figure 1.6. All programs and worksheets from this book can be found at http://www.mapleapps.com/powertools/EssentialMaple7/EssentialMaple7.shtml.

Now we test it.

```
>   restart:
```

We read programs in with the `read` statement. Note the Unix-style slashes in the path to this file, even though this is run on a Windows 2000 machine: Maple is originally a Unix product, and therefore the designers of Maple felt free to use the backslash for another purpose, namely, escape in strings. Therefore you must use the Unix style forward slash to indicate directories in the path. I typically name files after the main routine found in them, but there is no constraint by Maple that this should be so.

```
>   read "C:/books/ess/programs/FourierSineSeries.mpl";
```

Typing `FourierSineSeries` more than once is unnecessary, because we can use `macro` to define a shortcut for the current session.

```
#
# PROGRAM: FourierSineSeries:  compute the Fourier sine series of an
#                              input function.
#                              Interval normalized to [0,1].
#
# MAINTENANCE HISTORY:
#   Version 2:              May 23, 2001
#   First version           Feb 17 1994.
#   Modified to increase readability    May 25 1994.
# BASIC IDEA:    Standard theory of Fourier series.
# REFERENCE:     William E. Boyce and Richard C. DiPrima,
#                Elementary Differential Equations and
#                Boundary Value Problems, 2nd. ed., Wiley, 1969,
#                pp. 423--429.
# CALLING SEQUENCE: FourierSineSeries(fn, x);
# INPUT:  fn : an expression denoting a function of x, e.g.,
#              sin(x) or cos(x^3).
# OUTPUT: an  O P E R A T O R  which takes an integer n as
#         argument and returns a sum of n terms of the
#         Fourier sine series for fn.
# KNOWN BUGS/WEAKNESSES:  May occasionally divide by zero.
#                         See the exercises.
# Miscellaneous remarks:
#    This version uses nested lexical scopes and escaped local
#    variables to allow the program to work on more than one
#    function in the same session.  See Chapter 3.
FourierSineSeries := proc( f, x::name )
   local c, j, k;
   description "FSS(f,x): Compute the Fourier sine series of f wrt x";

   c := 2*int( f*sin(k*Pi*x), x=0..1);

   if hasfun(c, int) then
      error "Sorry, couldn't do the integral explicitly";
   else
      c := eval( c, [sin(k*Pi)=0, cos(k*Pi)=(-1)^k] );
      c := unapply( collect( c, k, factor ), k );
      # c(k) will divide by zero sometimes
      return n -> add( c(k)*sin(k*Pi*x), k=1..n );
   end if;
end proc:
```

Figure 1.6: A Maple program to compute the Fourier sine coefficients of a given function

```
>  macro( FSS=FourierSineSeries );
```
$$FSS$$

```
>  f := FSS( x^2*(1-x^2), x );
```
$$f := n \to \mathrm{add}(c(k)\sin(k\,\pi\,x),\, k = 1..n)$$

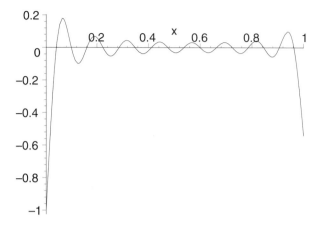

Figure 1.7: Graph of the error in the fifteen-term solution with a nonsmooth initial condition

> f5 := f(5);

$$f5 := \left(16 \frac{1}{\pi^3} - \frac{96}{\pi^5} \right) \sin(\pi x) - \frac{3 \sin(2 \pi x)}{\pi^3}$$

$$+ \left(\frac{16}{27} \frac{1}{\pi^3} - \frac{32}{81} \frac{1}{\pi^5} \right) \sin(3 \pi x) - \frac{3}{8} \frac{\sin(4 \pi x)}{\pi^3}$$

$$+ \left(\frac{16}{125} \frac{1}{\pi^3} - \frac{96}{3125} \frac{1}{\pi^5} \right) \sin(5 \pi x)$$

We see by inspection that this is the same result as before. Now we consider a different function.

> f := FSS(cos(x), x);

$$f := n \rightarrow \text{add}(c(k) \sin(k \pi x), k = 1..n)$$

> f15 := f(15):

> plot(f15-cos(x), x=0..1);

That plot is shown in Figure 1.7. We see the Gibbs phenomenon—the overshoot near the discontinuity (see [4])—clearly in that plot. The convergence is less good for this example because the *odd* extension (which of course is all that a Fourier sine series can converge to) of $\cos(x)$ to $-1 \leq x \leq 1$ has a jump discontinuity at the origin, and another at $x = 1$. See the exercises.

The program FourierSineSeries is intended neither as an example of elegant, robust programming nor for general-purpose use: It is an example of a "throwaway" program written quickly to solve one particular problem. I find

Maple extremely useful in a wide variety of problems and the creation of such "throwaway" programs to be a major help in day-to-day computation.

It is also easy to write much more powerful, robust, and general-purpose programs in the Maple programming language. For examples of such programs, you can look at the source code for the Maple library itself, or in [21]. The present book limits its scope to teaching the use of Maple as a calculator and the construction of small to moderately small programs.

A Shorter, More Robust Version

If we ask the program of the previous section for the Fourier sine series of $f(x) = \cos(\pi x)$, we get an error: "division by zero." Question 3 below asks you to explain why. The following simple program does not suffer from this problem.[8]

```
>   restart;
>   FourierSineSeries2 := proc( f::operator )
>      local k, z;
>      description "FSS2(f::operator):"
>            " improved Fourier sine series of f";
>      (n,x) -> add( limit( 2*int(sin(k*Pi*z)*f(z),
>                  z=0..1),
>                   k=j
>                  ) * sin(j*Pi*x),
>             j=1..n
>             )
>   end proc:
```

Note that Maple allows the convenient entry of long strings by simply juxtaposing the pieces; this is an implied concatenation (though the help file denies it).

```
>   macro( FSS2=FourierSineSeries2 );
                        FSS2
```

When we try it on our example from before, we get a reminder that we changed the input syntax:

```
>   FSS2( x^2*(1-x^2), x );

Error, invalid input: FourierSineSeries2 expects its
1st argument, f, to be of type operator, but received
x^2*(1-x^2)
```

Using the new input syntax, we get

---

[8]*Warning:* In order to display the dummy variable of integration differently from the variable of the procedure, I had to change the input to this routine to be an *operator*.

```
>    f := FSS2( x -> x^2*(1-x^2) );
```

$$f := (n, x) \rightarrow \text{add}($$

$$\left(\lim_{k \to j} 2 \int_0^1 \sin(k\, \pi\, z)\, (x \rightarrow x^2\, (1 - x^2))(z)\, dz \right)$$

$$\sin(j\, \pi\, x),\ j = 1..n)$$

I modified the display of the output slightly for this book. Note that the application of the operator $x \rightarrow x^2(1-x^2)$ to its argument z has not simplified automatically to $z^2(1-z^2)$. It will when the operator f is called with an actual argument. See Section 3.2 for a fuller discussion of operators.

```
>    f5 := f( 5, x );
```

$$f5 := \frac{16}{\pi^5}\, (-6 + \pi^2)\, \sin(\pi\, x) - \frac{3}{\pi^3}\, \sin(2\, \pi\, x)$$

$$+ \frac{16}{81\pi^5}\, (-2 + 3\, \pi^2)\, \sin(3\, \pi\, x) - \frac{3}{8\pi^3}\, \sin(4\, \pi\, x)$$

$$+ \frac{16}{3125\pi^5}\, (-6 + 25\, \pi^2)\, \sin(5\, \pi\, x)$$

This looks different than before, but is equivalent (and both are correct). But now if we try $\cos(\pi x)$, which gave an error with the previous program, we get

```
>    fc := FSS2( x -> cos(Pi*x) );
```

$$fc := (n, x) \rightarrow \text{add}($$

$$\left(\lim_{k \to j} 2 \int_0^1 \sin(k\, \pi\, z)\, (x \rightarrow \cos(\pi\, x))(z)\, dz \right) \sin(j\, \pi\, x),$$

$$j = 1..n)$$

```
>    fc( 10, x );
```

$$\frac{8}{3\pi}\, \sin(2\, \pi\, x) + \frac{16}{15\pi}\, \sin(4\, \pi\, x) + \frac{24}{35\pi}\, \sin(6\, \pi\, x)$$

$$+ \frac{32}{63\pi}\, \sin(8\, \pi\, x) + \frac{40}{99\pi}\, \sin(10\, \pi\, x)$$

which is correct. The function that makes this possible in this case is `limit`, a very powerful utility for taking limits in Maple. This power is not without price: Instead of simply evaluating each coefficient in the Fourier series, we are taking limits for each of the n terms; this can get expensive just to catch the one or two terms that cause problems with the other method. The goal of coping with all possible inputs sometimes imposes a larger computational cost than just coping with "some" inputs.

In Chapter 3 we revisit this example and construct a program that tries the simple method first, and only when that fails does it use `limit`.

Exercises

1. Use `piecewise` to plot the odd extension to $\cos x$ on $-1 \le x \le 1$:

$$y = \begin{cases} \cos x & x > 0 \\ 0 & x = 0 \\ -\cos x & x < 0 \end{cases}$$

Plot the 15-term Fourier sine series for this function on the same plot.

2. Solve $u_t = u_{xx}$ on $0 \le x \le 1$ for $t > 0$ given that $u(x, 0) = x^3(1 - x^3)$, $u(0, t) = u(1, t) = 0$.

3. If instead $u(x, 0) = x \sin \pi x$, the program gives a division by zero when computing, say, `f10`. Why? Will using `limit` work in this case? What about for $u(x, 0) = x \sin^2 \pi x$?

4. The procedure in Figure 1.6 computes a Fourier sine series of an odd function on the interval $[0, 1]$. Modify it so that it solves the heat conduction problem with initial condition $u(x, 0) = f(x)$ instead.

1.1.4   Arithmetic

Maple has facilities for arbitrary-precision integer and rational arithmetic (both real and complex), modular arithmetic, hardware floating-point and arbitrary-precision floating point arithmetic, and the arithmetic of matrices.

A venerable example of arbitrary-precision integer arithmetic is the computation of $3!!! = 6!! = 720!$, as seen in Figure 1.8. [It is often used because it fits nicely on the screen of a typical window.]

In 1994, when the first edition of this book was written, the calculation in Figure 1.8 already took less than a second, on a 25MHz 486 IBM PC clone. Today (July 2001), on my (slow) 233Mhz notebook, the measured time to compute this (0.003 seconds) is less than the probable error in that measurement. The length of the final answer usually provokes laughter, followed shortly thereafter by the question "Is it right?"

Let's explore that question, without further use of Maple. Stirling's approximate formula $n! \sim \sqrt{2\pi n} n^n \exp(-n)$ gives $720! \approx 2.60091 \cdot 10^{1746}$, and this agrees with the first three digits of the above. In the printed answer above we count the digits in the rows and multiply out, giving 1747 digits in all, which agrees with the magnitude of the result of Stirling's formula. Finally, we can count the number of factors of 5 in $720!$, which will give us the number of trailing zeros in the answer (because there will be more than enough factors of 2 to make each 5 a 10). The number of factors of 5 is

$$\left\lfloor \frac{720}{5} \right\rfloor + \left\lfloor \frac{720}{5^2} \right\rfloor + \left\lfloor \frac{720}{5^3} \right\rfloor + \left\lfloor \frac{720}{5^4} \right\rfloor + \left\lfloor \frac{720}{5^5} \right\rfloor + \cdots$$

```
>   3!!!;
```

$$
\begin{aligned}
&260121894356579510020490322708104361119152187501694\,57\backslash\\
&8572754183785083563115694738224067857795813045708261\,9\backslash\\
&9205758922472595366415651620520158737919845877408325\,2\backslash\\
&9105244690388811884123764341191951045505346658616243\,2\backslash\\
&7194019711390984553672727853709934562985558671936977\,4\backslash\\
&0700037004307837589974206767840169672078462806292290\,3\backslash\\
&2107161669867260548988445514257193985499448939594496\,0\backslash\\
&6404513236214026598619307324936977047760606768067017\,6\backslash\\
&4916694030348199618814556251955925669188308255149429\,4\backslash\\
&7596537274845624628824234526597789737740896466553992\,4\backslash\\
&3592878621251596748322097602950569669992728467056374\,7\backslash\\
&1375330192483135870761254126834158601294475660114554\,2\backslash\\
&0749589952563543068288634631084965650682771552996256\,7\backslash\\
&9084523570255218622235813001670083452344323682193579\,3\backslash\\
&1847019565107297818043541738905607274280485839959197\,2\backslash\\
&9021726612291298420516067579036232337699453964191475\,1\backslash\\
&7556755769539223380305682530859997744167578435281591\,3\backslash\\
&4613403946049012695420288383471013637338244845066600\,9\backslash\\
&3348484440711931292537694657354337375724772230181534\,0\backslash\\
&3264717753198453734147867432704845798378661870325740\,5\backslash\\
&9389242157096959946305575210632032634932092207383209\,2\backslash\\
&3356309923267504401701760572026010829288042335606643\,0\backslash\\
&8988871029738079757801305604957634283868305719066220\,5\backslash\\
&2911748225105366977566030295740433879834715185526028\,0\backslash\\
&5333866357139101046336419769097397432285994219837046\,9\backslash\\
&7910995630338960467588986579571117656667003915674815\,3\backslash\\
&1159439800436253993997312030664906013253113047190288\,9\backslash\\
&8491856203766669164468791125249193754425845895000311\,5\backslash\\
&6168297430464114253807489728172337595538066171980140\,4\backslash\\
&6779356147936352662656833395097600000000000000000000\,0\backslash\\
&0000000000000000000000000000000000000000000000000000\,0\backslash\\
&0000000000000000000000000000000000000000000000000000\,0\backslash\\
&0000000000000000000000000000000000000000000000000000
\end{aligned}
$$

Figure 1.8: A large integer computed in Maple

and since $5^5 = 3125$, this series terminates at the fourth term. The value of the sum is then $144 + 28 + 5 + 1 = 178$, and we see by counting that there are 178 zeros in the answer printed above. These simple (and partial) checks, independent of Maple's computations, boost our confidence that the answer Maple gives is correct.

There is a system-dependent size limitation on Maple integers, typically that they must have roughly fewer than 268,000,000 digits. You can query the limitation on your system by issuing the command `kernelopts( maxdigits )`. This limitation is rarely exceeded in practice.

Representing such large objects requires Maple to make extensive use of computer memory, and to manage it well. Memory management is done through what is known as a *garbage collector*, which periodically runs through the computer memory used in a session and frees up space to be reused. You do not need to explicitly manage memory in Maple, although you can adjust the frequency of garbage collection by calling gc. See ?gc for details.

Maple fractions are simply pairs of integers, kept relatively prime by automatic GCD computations. The data type `rational` includes integers and fractions.

Exercises

1. Find the exact number of possible bridge hands. [A full deck of 52 cards is dealt randomly to four people; a "hand of bridge" is the result of any such deal. The order of the cards in each hand is not relevant, but the people are seated in a definite order.]

2. How much computer memory does a FORTRAN (or C) single- or double-precision floating point number require under the IEEE standard [47]?

3. Explain the following statement and infer its proper context.

   "An n-by-n matrix requires $O(n^2)$ storage and $O(n^3)$ operations to invert."

   In particular, suggest circumstances when it will not be true in Maple. Ignore improvements to Gaussian elimination such as Strassen's algorithm; the reason that this is not true in Maple isn't because Maple can do *better*, but rather the opposite. Be concise but clear.

4. Use Newton's iteration $x_{n+1} = x_n - f(x_n)/f'(x_n)$ to generate five rational approximations to the root x^* near $x_0 = -1$ of

$$f(x) = x^3 - \frac{1}{10}x + 1 = 0.$$

   Since the error $e_n = x_n - x^*$ behaves approximately as $e_{n+1} \propto e_n^2$, estimate the error in your *best* approximation. What happens if you give a symbolic initial guess?

5. Evaluate, directly and also by using logarithms,

$$C_3^{10^{30}} \cdot \left(\frac{1}{10^{24}}\right)^3 \left(1 - \frac{1}{10^{24}}\right)^{10^{30}-3}$$

to 60 places. Note: $C_m^n = n!/(m!(n-m)!)$ is the binomial coefficient. This number is the probability that exactly 3 atoms out of a sample of 10^{30} decay in a period where the probability of any single atom's decay is 10^{-24}.

6. The Maple `convert` procedure has many uses. Convert 55/89 into "continued fraction form," e.g.,

$$\frac{339}{284} = 1 + \cfrac{1}{5 + \cfrac{1}{6 + \frac{1}{9}}} ,$$

using the list of partial quotients notation, e.g., [1, 5, 6, 9]. See `numtheory[` `cfrac ]` for an alternative.

1.1.5 Interrupting a Maple Computation

Maple will sometimes "go away" for quite a while to do its calculations. You may wish to interrupt the calculation rather than wait for Maple to finish. This can usually be done, although since Maple makes complex use of memory it is not always possible to stop computation instantly. For some command-line Maple systems, pressing CTRL-Break will interrupt the calculation; on other systems it is CTRL-C, and on Unix it is the Unix interrupt character (for many users, *break*). For windowing systems, click on the interrupt button (marked with a Stop sign) in the toolbar at the top of the window. As an exercise, execute the following command, and interrupt it before it finishes.

```
>   int( 1/(t^1000-1), t );
```

```
Warning, computation interrupted
```

Check your local documentation for the precise interrupt key if these do not work, but be aware that Maple will not always respond immediately to an interrupt.

1.1.6 Saving Work

Save your work with the `save` command, which saves your variables and procedures. In the following example session several typical Maple commands are issued, followed at the end by two `save` statements that demonstrate the main ways this command is used. The sample Maple session itself should be reasonably intelligible even though you may not know the meaning of the commands used, so just skim it and pay particular attention only to the `save` statements.

```
>   restart:
```

```
>   with( LinearAlgebra ):
```

```
>   A := RandomMatrix( 3, 3 );
```

$$A := \begin{bmatrix} -21 & -50 & -79 \\ -56 & 30 & -71 \\ -8 & 62 & 28 \end{bmatrix}$$

```
>  p := CharacteristicPolynomial( Transpose(A).A, x );
```

$$p := -1478094916 + 114521881\,x - 22951\,x^2 + x^3$$

```
>  rts := [ fsolve( p, x, complex ) ];
```

$$rts := [12.94019843, 7308.069806, 15629.99000]$$

```
>  rootBounds := realroot( p, 1/10^8 );
```

$$rootBounds := \left[\left[\frac{1736804033}{134217728}, \frac{868402017}{67108864}\right],\right.$$

$$\left[\frac{980872525423}{134217728}, \frac{61304532839}{8388608}\right],$$

$$\left.\left[\frac{1048910872935}{67108864}, \frac{2097821745871}{134217728}\right]\right]$$

```
>  evalf(rootBounds,15);
```

$$[[12.9401984289289, 12.9401984363794],$$
$$[7308.06980597228, 7308.06980597973],$$
$$[15629.9899955839, 15629.9899955913]]$$

```
>  save A, p, 'mymatrix.mpl';
```

```
>  save p, rts, rootBounds, 'myroots.m';
```

See the help entries for LinearAlgebra, fsolve, realroot, *etc.*, for more information on what those commands are capable of.

The command

```
>  save A, p, 'mymatrix.mpl';
```

saves only the matrix A and the polynomial p in the human-and-Maple-readable file mymatrix.mpl. Please be aware that there are some known problems with that form of the command, particularly with saving series or tables. The command save p, rts, rootBounds, 'myroots.m' saves the polynomial, its approximate roots, and their root bounds, into the Maple-readable file myroots.m. The file extension tells Maple which format to use. The .m format is not human-readable but is more efficient for Maple to read. The left quotation marks (') are different from right quotation marks (') and are necessary for the file names so that the Maple *noncommutative multiplication operator* period (.) doesn't act on the file names. Many people consistently use the file extension .mpl to indicate a Maple input file. This convention is, of course, a preference, and not a Maple requirement, so long as you reserve .m for the Maple-readable-only files.

1.2 Some Things to Watch Out For

Your previous experience with computers may have taught you to expect certain patterns; some of these can lead you astray in Maple. In particular, variables can have a much wider variety of values in Maple than they can in, say, FORTRAN or C.

1.2.1 Good Worksheet Hygiene

Worksheets have become the dominant method of using Maple interactively. The following simple rules will help you to write easy-to-read, maintainable, and correct worksheets.

1. Start your worksheets with a `restart` command. This gives more consistency when you run them more than once.

2. Worksheets allow you to execute Maple commands out of "top-down" sequence. *Don't use this feature.* Always maintain a top-down execution order for your worksheets.[9]

3. You can write small Maple procedures in a worksheet, and they might even be readable if you use shift-return to break lines, and use white space. Don't do this for programs that are more than about ten lines long. Instead, use your favourite editor to create and maintain your Maple programs, and use the `read` command or place them in a Maple repository (see `?repository`). Waterloo Maple Inc. recommends the VIM editor, available at www.vim.org, or emacs with its Maple mode. I use WinEdt on Windows systems, available at www.winedt.com (this is not WinEdit), and emacs on Unix systems.

4. Use the title, section, and subsection features (from the Insert menu). Collapsing sections hides information in a useful way.

5. Use text often (F5 turns a Maple line into a text line) to explain what you are doing before each step, or to explain what the result means (or both).

6. Turn on the "phrase" option for assumed variables, because the trailing tildes are often mistaken for minus signs. You can do this from the Options/Assumed Variables menu or by issuing the command `interface( showassumed=2 )`. See the discussion in Section 1.7.

7. Use the long forms for many package routines, e.g., `plots[display]` rather than loading in the whole package. This avoids some spurious redefinitions of names, and the redundancy helps comprehensibility.

---

[9]Like traffic laws in hot-blooded countries, this rule can be considered just a suggestion.

8. Be aware of possible "ordering problems." Maple does not guarantee that it will produce sets or sequences of answers from its commands in the same order from one session to the next. For example, one time your intermediate result may be $\{x, y\}$, and $\{y, x\}$ the next.[10] Therefore, selecting the first element of a list or set and proceeding with the computation can sometimes cause unexpected behaviour when you execute the worksheet a second time. Instead of selecting via a numerical index, try to select entries from a list by their characteristics, using `select`. This will tend to produce a worksheet that will more likely give reproducible results.

1.2.2   Common Syntax Errors

Like all computer languages, Maple demands a certain amount of precision from its users. This is sometimes annoying, since no one likes to get syntax error messages. However, these messages are useful, once you learn what they mean. The following session contains examples of most of the common syntax errors, and the messages that result.

Forgetting the multiplication symbol is a common error.

```
>   2x+3;
```

```
Error, missing operator or ';'
```

On a windowing system, the flashing bar cursor is placed between the 2 and the x, where Maple thinks the error is.

Typing the wrong number of parentheses (or other fences) is also a syntax error.

```
>   sin(exp(x);
```

```
Error, ';' unexpected
```

Using the wrong sort of quotation marks (see Section 1.2.8) can cause difficulty:

```
>   'Not a string, since the quotation marks are
>   wrong';
```

```
Error, missing operator or ';'
```

Splitting an indivisible object across lines without using the continuation character (\ ) causes a syntax error.

```
>   1234
```

```
>   5678;
```

```
Error, unexpected number
```

Instead, that should have been the following.

---

[10]This "feature," which arises because Maple hashes objects and expressions in address order, for "efficiency," requires awareness on the part of the user.

```
>   1234\
>   5678;
```

$$12345678$$

Some 'syntax errors' produce legal Maple code with a different meaning from that intended. Since Maple does not complain about this type of 'syntax error', they can be much harder to find. For example, incorrect capitalization will produce legal Maple code, as in the following example.

```
>   SIN(x);
```

$$SIN(x)$$

```
>   diff( SIN(x), x );
```

$$\frac{\partial}{\partial x} SIN(x)$$

Unless the user has previously defined his or her own function SIN, she or he probably meant for Maple to differentiate $\sin x$ to get $\cos x$; the above result might convince him or her that Maple was stupid. In fact, Maple's name for the sin function is sin, not SIN. But an error message would be inappropriate for the above session, since the user might really want to talk about some other function called SIN. As an exercise, what do you predict Maple will do if you type

```
>   sin x;
```

or

```
>   sinx;
```

at the command prompt?

1.2.3   Assigning Values to Variables

First, if an identifier (say x) has not been assigned a value, then in Maple it stands for itself; that is, it is a *symbol*. This is different from FORTRAN, where reference to an undefined variable may get you a "Not a Number" (NaN), zero, an error message, or garbage, depending on the compiler. It is also different from Java, which guarantees to preinitialize all variables to clean default values (zero for numeric ones).

In Maple, symbols (variables) are assigned values with := (pronounced "becomes") as in Pascal and *not* with = as in C or FORTRAN. It is common to confuse the two.

```
>   p := x^2 + 3*x + 7;
```

$$p := x^2 + 3x + 7$$

```
>   x;
```

$$x$$

Note that x has no value (it represents itself) and that p has been assigned a value.

> p;
$$x^2 + 3x + 7$$

> q = p;
$$q = x^2 + 3x + 7$$

That looks like assignment, but it is an *equation*.

> q;
$$q$$

Nothing happened to q as a result of the previous statement. You can assign equations to variables:

> r := q=p;
$$r := q = x^2 + 3x + 7$$

The variable r has the value $q = x^2 + 3x + 7$. This kind of value for a variable is not possible in a purely numerical language such as FORTRAN.

1.2.4 Removing Values from Variables

A candidate for the most common cause of bugs is *regarding a variable that has a value as a symbol*. For example, suppose that in the first part of your session, you set x := 3, and much later you define p := x^2. Then suppose you try to integrate p with respect to x, having forgotten that x has a value; you wind up asking Maple to integrate with respect to 3, which doesn't make sense. We will see an example of this shortly. You can 'unassign' variables by a statement of the form x := 'x'. See also the command unassign.

> p := x^2+sin(x);
$$p := x^2 + \sin(x)$$

> x := 3;
$$x := 3$$

Now suppose we do other things for a while, and forget that $x = 3$. Now we want to integrate p.

> int(p, x);

Error, (in int) wrong number (or type) of arguments

This error message is not terribly illuminating to the uninitiated.

> x := 'x';
$$x := x$$

> int(p, x) + C;
$$\frac{1}{3}x^3 - \cos(x) + C$$

Order is important up there: let's try assigning x, then p.

```
>   x := 3;
```

$$x := 3$$

```
>   p := x^2+sin(x);
```

$$p := 9 + \sin(3)$$

```
>   int( p, x );
```

```
Error, (in int) wrong number (or type) of arguments
```

Again, we must interpret that error message (it means the same as it did last time; we are trying to get Maple to integrate with respect to 3).

```
>   x := 'x';
```

$$x := x$$

```
>   int( p, x )+C;
```

$$(9 + \sin(3))\, x + C$$

Now what happened?

In the last part we assigned a value to p after we had assigned a value to x. Full evaluation of the right-hand side takes place, so in the second part, p was assigned the value $9 + \sin(3)$, with no more links or references to x. Hence the computed answer.

1.2.5   `sign` **versus** `signum` **versus** `csgn`

The functions `sign`, `signum`, and `csgn` are often confused, because different people mean different things by the "sign" function. In Maple the function `sign` is intended to compute the signum of the *leading coefficient of a polynomial*. It is a common error to use `sign` when you mean `signum`. The definition of `signum` is complex-valued for complex arguments:

$$\operatorname{signum}\left(re^{i\theta} \right) := e^{i\theta} \tag{1.1}$$

if $r > 0$ and *undefined* if $r = 0$. In particular, if $z > 0$, then signum(z) is 1, and if $z < 0$, then signum(z) is -1. See ?signum for details of choosing the value you would like to have at the origin; note that there are many different standards in use, and applications where each is more useful than any other.

The function `csgn` is real-valued for complex arguments. It is best defined in terms of the unwinding number $\mathcal{K}(z)$, about which more can be found in Appendix A, or in [10]; here is one definition of $\mathcal{K}(z)$:

$$\mathcal{K}(z) := \left\lceil \frac{\Im(z) - \pi}{2\pi} \right\rceil, \tag{1.2}$$

where $\Im(z)$ is the imaginary part of z. This arises in simplifying complex logarithms:

$$z = \ln e^z + 2\pi i \mathcal{K}(z) \,. \tag{1.3}$$

In terms of this function,

$$\mathrm{csgn}(z) := \exp\left(\pi i \mathcal{K}(2\ln z)\right) \,. \tag{1.4}$$

You may compare this to the detailed definition in ?csgn. Clearly, the value of csgn is either $+1$ or -1 and is ambiguous at $z = 0$. It most often occurs in simplification of square roots: $\sqrt{z^2} = \mathrm{csgn}(z)z$. A generalization that allows simplification of nth roots is $C_n(z) := \exp(2\pi i \mathcal{K}(n\ln z)/n)$.

1.2.6 Accidental Creation of a Remember Table

```
>  restart;
```

People often try to create operators or procedures by a statement of the form

```
>  p(x) := sin(x) + exp(x);
```

$$p(x) := \sin(x) + e^x$$

and are very surprised that it doesn't work:

```
>  p(1);
```

$$p(1)$$

even though $p(x)$ has the value they expect:

```
>  p(x);
```

$$\sin(x) + e^x$$

but not if the argument to p is anything except x.

```
>  p(a);
```

$$p(a)$$

What has happened is that the original assignment to $p(x)$ created a procedure called p, and placed an entry in its "remember table" (see Chapter 3) so that if p is called with exactly the argument x, then it "remembers" the expression it was told. Further assignments like p(y) := 7 or p(blue) := 13 would tell the procedure what to do with arguments y or blue but in no case give a general rule. Let us look at the procedure defined by just the assignment p(x) := sin(x)+exp(x); above:

```
>  eval( p );
```

proc() **option** *remember*; 'procname(args)' **end proc**

To look at the remember table for a function you can issue the command

```
>  op( 4, op(p) );
```

$$\text{table}([x = \sin(x) + e^x])$$

Probably what was intended was the creation of an *operator*:

```
>  p := x -> sin(x) + exp(x);
```

$$p := x \rightarrow \sin(x) + e^x$$

```
>  p(1);
```

$$\sin(1) + e$$

```
>  p(x);
```

$$\sin(x) + e^x$$

```
>  p(a);
```

$$\sin(a) + e^a$$

```
>  eval( p );
```

$$x \rightarrow \sin(x) + e^x$$

1.2.7   Fences: Parentheses ( ) versus Braces { } versus Brackets [ ] versus Angle Brackets ⟨ ⟩

In Maple, parentheses ( ) are used to surround arguments to procedures and to group expression sequences; for example, p(x,y) sends x and y to the procedure p, whereas (x,y) on its own is just the expression sequence x,y grouped together. Maple statements, such as read, error, or return, do not need their arguments grouped with parentheses (but it doesn't hurt).

Braces { } are used to delimit *sets*, that is, composite objects that contain unique objects: writing {x,y,x} results in {x,y} as output (or perhaps {y,x}; order is unimportant to a set, and cannot be relied upon).

Brackets [ ] are used to delimit *lists*, e.g., [$a1, a2, a1, a3$], which are ordered composite objects (expression sequences) that may contain duplicate entries. Brackets are also used to delimit indices into tables, arrays, vectors, matrices, Vectors, and Matrices. An example is A[i,j].

Angle brackets ⟨ ⟩ are used as a "shortcut" to construct matrices. To construct an object with three columns type < a | b | c >; to construct an object with three rows type < a, b, c >. If the elements a, b, and c are compatible Vectors then the result is a Matrix. This shortcut is not used in this book because I prefer the Matrix and Vector constructors.

1.2.8   Quotation marks: Left versus Right versus String

The distinction between the different types of quotation marks is important, and many more examples of their use will be given in this book. To get help from Maple on the different quotation marks, type ?quotes.

Left quotation marks are name delimiters.

```
>   `This is a complicated name with spaces`:
```

Right quotation marks delay evaluation.

```
>   'sin(Pi)';
```

$$\sin(\pi)$$

```
>   %;
```

$$0$$

Double quotation marks (") delimit strings:

```
>   "This is a string";
```

$$\text{``This is a string''}$$

Earlier versions of Maple used " (the same character as now delimits strings) to refer to the previous result, and this function has now been taken over by the percent sign %. Similarly, double percents (%%) refer to the result before that, and triple percents (%%%) refer to the result before that. If you wish to refer to still earlier results, you must have previously issued the history command or else named the results in an assignment statement. Otherwise, you are out of luck. In worksheets, you can always move the cursor up and reexecute commands (this will alter % and so on). It is also often possible to reexecute input commands using the 'history' mechanism of your operating system; for example, on some platforms, in command-line Maple simply hitting the up-arrow will recover the previous command, and hitting it again will get the one before that, and so on. Reexecuting commands is usually less efficient than saving the results the first time, but also may take less time than the first computation did because many Maple commands *remember* their previous workings.

See Section 3.4 for more discussion on the percent variables.

See ?string for a discussion on strings in Maple. Complicated strings can be formed from other strings by using the double bar (||) operator, which is the Maple concatenation operator. A common example of this is the creation of several related names a1, a2, etc., although this may be more usefully done with indexed names, and more safely with `tools/genglobal`. The following introduces the notion of a *range* in Maple. A range is simply two integers connected with two periods, as follows, and it means all the integers from the lower to the upper value, inclusive (except, as noted before, when used in int, when it delimits a real interval).

```
>   a||(1..5);
```

$$a1,\ a2,\ a3,\ a4,\ a5$$

```
>   p := add( b||i * x^i, i=0..5 );
```

$$p := b0 + b1\,x + b2\,x^2 + b3\,x^3 + b4\,x^4 + b5\,x^5$$

That is unsafe, because perhaps you have already defined b3, for example. You can guarantee that things are ok (i.e., generate unique, unused names) by using `tools/genglobal`, as follows.

```
>  restart;
```

We first ensure that the name *b* itself is free:

```
>  'tools/genglobal'(b);
```

$$b$$

```
>  for i from 0 to 5 do b[i]:='tools/genglobal'(b);
>  end do;
```

$$b_0 := b0$$
$$b_1 := b1$$
$$b_2 := b2$$
$$b_3 := b3$$
$$b_4 := b4$$
$$b_5 := b5$$

```
>  add( b[k]*x^k, k=0..5 );
```

$$b0 + b1\,x + b2\,x^2 + b3\,x^3 + b4\,x^4 + b5\,x^5$$

In this case all of the symbols from b1 to b5 were free. If we had assigned b2 to something before executing this loop, then genglobal would have skipped over b2.

Any Maple compound symbol (a nonindexed name, delimited by left quotation marks) can be used as a variable name. This can sometimes look a little strange, and so I recommend that only simple strings be used as variable names, following common practice. Here is a "strange" example, to show what I mean.

```
>  '2+2' := 5;
```

$$2 + 2 := 5$$

1.2.9 Precedence of Operators

The Maple arithmetic operators are + for addition, * for multiplication, - for subtraction and negation, / for division, mod for modular arithmetic, and ^ for exponentiation. Note that you must parenthesize a^b^c explicitly, either as a^(b^c) or (a^b)^c, whichever you really mean; if you leave off the parentheses, Maple will generate a syntax error.

The precedence of these operations is the natural one, where exponentiation takes place first, followed by multiplication and division, and finally by addition and subtraction. Arithmetic expressions are read left to right, subject to parentheses and the above precedence rules, as is standard in most computer languages. Thus 1 + 2*3^3 produces 55 and not 729.

See ?precedence for a complete list of the precedence hierarchy.

```
!, $, *, +, -, ., /, <, =, >, @, D, O, ^
```

Figure 1.9: The one-character names that are protected in Maple

1.2.10   Protected and Reserved Names

You may not use Maple reserved words as variable names: for example, `error`, `try`, `if`, `fi`, `do`, `od`, `next`, and other control structure names. Maple will immediately object to the use of reserved words as variable names, with syntax errors.

Some other names are not "reserved" as part of the Maple programming language, but are "protected". These are usually names of crucial pieces of library or kernel code, such as `op`, `Pi`, and `D`. You *can* undo this protection by issuing, e.g., the command `unprotect(Pi);`, but this is not recommended. See `?unprotect` for more details.

As an alternative, one can use `alias` to allow one to use a protected name:

```
>    alias( gamma=ggamma );
```
$$\gamma$$

(or instead of `ggamma`, some other unused name) allows you to type `gamma` and have it echoed and displayed as γ. This alias also allows Maple to generate the symbol `gamma` as the result of a computation. You can assign to it, but this really assigns to the variable `ggamma`, which is not protected or used elsewhere in Maple. The Maple routines that use Euler's constant `gamma` are not disturbed by this.

This discussion skirts a larger issue. I consider the `protect` command a step forward from the previous state, but I think that the Maple user should be allowed to use whatever single-letter names she or he wishes, for whatever purpose: Currently, the single-letter names D, I, `Pi`, `gamma`, `GAMMA`, `Zeta`, and many others[11] are forbidden to the user, to varying degrees. There is ongoing debate within the Maple group on how best to resolve this issue.

In Figure 1.1 we see some of the exceptions to the rules for using single Greek or Latin letters in Maple. The general rules are that typing the name of a Greek letter all in lowercase (for example, `xi`) causes it to pretty-print in a Greek font (for example, ξ), and that typing the name of a Greek letter with the first letter of the name capitalized gets you the pretty-printed letter in a capital Greek font. The exceptions in the table arose through historical accident, as Maple grew, by the efforts of different designers. Items marked with a † are particularly troublesome because unexpected and are sources of common errors.

The one- and two-character names that are protected in Maple are shown in Figures 1.9–1.10. Special characters must be enclosed in left quotation marks

---

[11] See Table 1.1; *of course* these are all single-letter names (just not all in the English alphabet).

Name	displays as	Meaning in Maple
beta	β	none
Beta	B	the Beta function (protected)
gamma	γ	Euler's constant (protected)
Gamma	Γ	none[†]
psi	ψ	none
Psi	Ψ	the ψ function (protected)
Chi	Chi	the hyperbolic cosine integral (protected)
xi	ξ	none
pi	π	none[†]
Pi	π	ratio of circumference to diameter (protected)
zeta	ζ	none[†]
Zeta	ζ	the Riemann ζ function (protected)
GAMMA	Γ	the Γ (gamma) function (protected)
ZETA	Z	none
PI	Π	none
CHI	X	none
D	D	differentiation operator (protected)
I	I	$\sqrt{-1}$, protected another way
O	O	Order symbol for series (protected)
e	e	none

Table 1.1: Some single-letter names and their meanings in Maple

```
%?, **, .., ::, <=, <>, >=, @@, Ci, Ei, GF,
Im, Li, Pi, Re, Si, gc, if, is, ln, op, or, ||
```

Figure 1.10: The two-character names that are protected in Maple

```
..., Add, BOX, Chi, Det, FFT, Gcd, HSV, HUE, Int, Lcm, Non,
One, Psi, Quo, RGB, Rem, Shi, Ssi, Sum, Svd, abs, add, and,
cat, cos, cot, csc, erf, exp, gcd, has, int, lcm, lhs, log,
map, max, min, mul, not, odd, quo, rem, rhs, sec, seq, set,
sin, sum, tan, xor, zip
```

Figure 1.11: The three-character names that are protected in Maple

(e.g. '%?') to be used as names. An explanation of why these are protected may be available in the help system (but is not in some cases). The low-level routine op is particularly important in Maple: it selects the desired operand of an expression, and is used throughout the Maple library. The routine gc, which performs garbage collection, is at least as important.

```
Beta, Diff, Eval, FAIL, GRID, LINE, MESH, MOLS, NONE, NULL, Norm,
PLOT, Prem, TEXT, VIEW, ZHUE, Zero, Zeta, _xml, ansi, bind, ceil,
copy, cosh, coth, csch, csgn, diff, done, echo, erfc, erfi, eval,
even, feof, frac, frem, goto, heap, help, iFFT, igcd, ilcm, ilog,
info, iquo, irem, list, map2, modp, mods, name, nops, norm, open,
plot, prem, quit, rand, real, root, sech, sign, sinh, sort, sqrt,
stop, subs, surd, tanh, thaw, time, trig, true, type, with
```

Figure 1.12: The four-character names that are protected in Maple

The three-letter names that are protected in Maple are shown in Figure 1.11. The names HUE and HSV are used in color options to plot. The other entries in that list, except One, have associated help files and can be looked up. In addition to One, the name of the number Zero is also protected in Maple 7. The three-letter word mod immediately simplifies to 'modp', which is one of the four-letter protected names (see Figure 1.12).

See also ?keywords for a list of reserved words, including try and use.

1.2.11   Having Different Assumptions about Domains

Many scientific programmers expect that all variables that begin with any of the letters I through N will be integers, while the rest are real. More perniciously, many people think that x is real but z is complex; ε is small and positive, while N is a large integer.

These, of course, are not Maple's default assumptions. To all routines except evalc, names might be complex (even x, a, or E). Calling evalc allows it to act on its default assumption that names (even z) are real. Some routines even allow for the possibility that a name, say t or λ, might be a matrix.

This observation exposes a further difficulty: Because different people have worked on different sections of Maple (since a long time ago), and because Maple is not a strongly typed language, some of the pieces of Maple have distinct sets of implicit assumptions. The moral of this is that if you know your variables are integers, positive, or whatever, then tell Maple so explicitly, unless you know that Maple will take things the right way.

Personally, I am very grateful that I don't have to declare the types of variables in Maple before I can use them, but the trade-off between interactive convenience and software-engineering-level reliability is a delicate one. "That's not a bug, it's a feature."

1.3   Documenting Your Work

Choose your variable and procedure names carefully. They should be neither too short (= too cryptic) nor too long (= too unwieldy). They should instead be *appropriate*. Calling a list of constants Fred after your friend might be amusing but calling it `List_of_Stirling_numbers` is much more understandable. Of course, that is hard to type, so `listirl` might be better still (though with better editors and macros there is less excuse for such laziness today; `List_of_Stirling_numbers` is clearly to be preferred for infrequent use). Use longer names for those very rare occasions when you have to use global variables, so as to minimize conflicts with other pieces of code. Use local and environment variables where possible. [See Section 3.4 for definitions of these.] Use `macro` and `alias` to allow short names for interactive use.

Pick a consistent programming indentation style for your code.[12] Reading your own programs two months later, you will be grateful. In Maple, there may be many ways of expressing the same action, but pick one and use it consistently.

Comment your work. The comment character in Maple is `#`. Anything after this character on a line is unseen by Maple. Useful comments tell how to use a particular piece of code, what is expected as input, and what the output will look like. Next most useful is a maintenance history. This lets you compare two copies of code and use the most recent. Many hard disks come to resemble the shells of ancient marine creatures, with old, fossilized bits of code and data hanging about. A maintenance history helps deal with this. Finally, comments giving references to the papers or books from whence the algorithms came, or discussing how the algorithms work, are also worthwhile.

For a procedure, write a help file for it also. This doesn't have to be large and can simply be the usage information mentioned earlier, together with a reference and an example.

For a module, write a help file for the module and a help file for each exported user-level routine.

For a large collection of files meant to be read in and executed in a particular way, write a README file describing the overall structure and purpose.

Of course, all this increases your workload, but greatly improves your total throughput and productivity, because much less time is wasted reinventing your own wheels.

Consider the example program in Figure 1.13, which is useful in proving numerical quadrature of black-box functions (that is, functions that can only be evaluated, not examined analytically) to be impossible [35]. Observe the following points:

1. Names are not too long (`spy`, `secrets`) but are intelligible. Use of single-letter names is appropriate for mathematical variables or symbols.

2. The procedure checks its input, both in the formal parameter list (see Chapter 3) and afterwards to see whether the names have values. The syntax

---

[12]I like 3-space indentation. However, Maple's output in the Export to LaTeX, which I have used for this book, uses two different indentation styles. I have tried to make this consistent, but I may have missed some.

```
#
# Spy function
# (c) Robert M. Corless 2001
#
# Used in a proof that numerical quadrature of black-box functions
# by point methods is impossible.
#
# Reference: William M. Kahan, "Handheld calculator evaluates integrals",
#            Hewlett-Packard Journal (31) 8 Aug 1980 pp. 23--32.

# The "spy" function returns 0, no matter what its numeric input is.
# As a side effect, it puts the sequence of points it is called with,
# in reverse order, in an expression sequence stored in the global
# variable "secrets", which is initialized to NULL when spy is loaded.

secrets := NULL:

spy := proc( x :: {name,numeric} )
    option 'Copyright (c) Robert M. Corless 2001';
    description "Spy on numerical quadrature.";
    global secrets;
    # We use the usual trick to delay evaluation in the call to
    # int; if we are called with a symbol, we just hold off doing anything.
    if type(x,name) then
       return 'procname'(args)
    else
       secrets := x,secrets; # Inefficient for large lists.
       return 0;
    end if;
end proc;


Malicious := proc( x :: {name, numeric} )
    local k, c;
    option 'Copyright (c) Robert M. Corless 2001';
    description "Sabotage numerical quadrature.";
    global secrets;
    c := 1.0e100;  # Could be anything.
    c*mul( (x-secrets[k])^2, k=1..nops([secrets]) );
end proc;
```

Figure 1.13: The Maple programs used in a proof that black-box quadrature is impossible

::{name,numeric} means that the parameter can be either a Maple name or a real numeric value.

3. "Option Copyright" prevents automatic printing of procedure bodies (see ?procedure and ?interface for the verboseproc option). It allows the printing of the "description" field, though, and this can be useful.

4. The comments take roughly as much space as the program. Now, that's partly because I am a wordy person, and it is easier for me to write a lot than to write a little. It is possible to boil the comments down to a bare minimum, and indeed this can be a useful exercise. But wordiness doesn't hurt, here, and the information given can be helpful to explain the code to others, or to yourself after enough time has passed.

5. The comments include some historical information. Typically, this would include a reference for the algorithm (in this case, the proof).

6. White space (blank lines and spaces) and consistent indentation are used to make the code easier to read.

7. The calling sequence, the expected input, the expected output, and any side effects are the most important items to document.

Here is the program in use:

```
>   restart;
```

We have to force the use of Maple software floats here, to avoid an issue of binary-to-decimal-to-binary conversion. To find out how many digits are used in the hardware floats on your system, issue the command `trunc( evalhf(Digits) );`.

```
>   Digits := trunc( evalhf(Digits) ) + 1;
```
$$Digits := 15$$
```
>   read "D:/books/ess/programs/spy.mpl";
```

$$spy := \mathbf{proc}(x::\{name,\ numeric\})$$
description "Spy on numerical quadrature."
$$\cdots$$
end proc

$$Malicious := \mathbf{proc}(x::\{name,\ numeric\})$$
description "Sabotage numerical quadrature."
$$\cdots$$
end proc

Because the source code for those programs terminated with a semicolon (;), not a colon (:), the procedure bodies printed; but because the procedures had "Option Copyright," the interior of the bodies was hidden. This hiding isn't much of a security feature, because it can be overruled by use of the `verboseproc` keyword in the interface command (see `?interface`). But it is sometimes convenient.

```
>   evalf( Int( spy, 0..1 ) );
```
$$0.$$

Now that the spy function has done its work, we can force Maple's numerical integrator to compute an incorrect answer.

```
>  evalf( Int( Malicious, 0..1 ) );
```
$$0.$$

That result is incorrect; it was bound to be, since int uses a consistent set of evaluation points. This proves that numerical quadrature of black-box functions cannot be done, even for smooth functions (of course, it cannot be done for discontinuous ones, for example a function that is zero at each member of secrets and one everywhere else). The following shows the choice of evaluation points int made on my machine and gives the true answer (which, by choice of c in Malicious, can be made to be any number at all).

```
>  Malicious(x);
```

$$.10\,10^{101}(x - .93900889217855136934)^2$$
$$(x - .06099110782144 8630664)^2$$
$$(x - .99928964018514642149)^2$$
$$(x - .00071035981485357851197)^2$$
$$\cdot\;(x - .999999999999999854)^2$$
$$(x - .146449376371092630\,10^{-15})^2$$
$$(x - .500000000000000000)^2\,(x - .99)^2\,(x - .618)^2\,(x - .5)^2$$
$$(x - .01)^2$$

There appear to be 20 digits in each of those numerical coefficients. Maple will sometimes increase precision to guard against rounding errors, but even if it doesn't (when it uses hardware floats, for example) it may print more digits than you expect. If you are using hardware floats, with 14 or 15 digits of precision depending on your system, Maple may print as many as 18 digits. This is because Maple uses base-10 floats, whereas IEEE standard hardware floats use base-2, and it can be proved that in order to reliably convert between Maple floats and IEEE double precision, 18 decimal digits suffice [23].[13]

```
>  seq( Malicious(a), a=[secrets] );
```
$$0., 0., 0., 0., 0., 0., 0., 0., 0., 0., 0.$$

Now do the integral symbolically. This gets the correct answer.

```
>  int( Malicious(x), x=0..1 );
```
$$.602110076391606\,10^{90}$$

---

[13]However, while that's sufficient in theory, it may not be in practice: you also need correct programs! I suspect that a subtle bug in exactly this conversion at different settings of Digits is what causes the spy and Malicious programs to fail in Maple if we use hardware floats. This will be investigated after this book is sent to Springer.

1.4 The Three Levels of Maple "Black Boxes"

Maple provides many "black boxes" for use in mathematical applications. Some of these are built-in and available immediately, such as `series` and `int`. We will refer to such routines as "main" routines. The first of the above examples, `series`, is written in C and is part of the Maple "kernel." The second of these is written in the Maple programming language and is part of the built-in library.

The third type of "black box" available in Maple comes in a 'package', such as the `LinearAlgebra` module. A module is a good way of implementing a package, which is a collection of more-or-less related routines that can be made accessible by using the `with` command, or whose routines can be accessed with a longer form without loading all the package routines. For example, in the following session we load the `student` package and use one of its routines.

```
>  restart;

>  with( student ):

>  f := x*sin( 3*arcsin(x) );
```
$$f := x \sin(3 \arcsin(x))$$

```
>  F := Int( f, x=0..1 );
```
$$F := \int_0^1 x \sin(3 \arcsin(x)) \, dx$$

It is a weakness in Maple 7 that this integral is not computed without help. If we convert the arcsin to its logarithmic form, the sine to its exponential form, and the square roots to their `RootOf` form, then Maple gets the integral; so one expects that a future version of Maple will get this integral without help from the user. Proving that $\sin(3 \arcsin(x)) = x(3 - 4x^2)$ for all complex x for any consistent choice of branch of arcsin and for any choice of closures on the branches seems complicated, though, and this may excuse Maple's weakness here.

```
>  value( F );
```
$$\int_0^1 x \sin(3 \arcsin(x)) \, dx$$

```
>  Fn := changevar( u=arcsin(x), F, u );
```
$$Fn := \int_0^{1/2 \pi} \sin(u) \sin(3\,u) \sqrt{1 - \sin(u)^2} \, du$$

```
>  value( Fn );
```
$$\frac{1}{5}$$

```
>  evalf( F );
```
$$.2000000000$$

We see that the computed value of the integral is correct. However, the user should be warned that `changevar` paid no attention to the legality of the change of variable.[14] This brings up an extremely important point.

1.5 No Nontrivial Software Package is Bug-Free

Maple has bugs. It has *always* had bugs. Though it is an evolving system and bug-fixing is a major activity of the Maple group, it always *will* have bugs. The same statements hold for *any* computer language, and indeed for any program I have ever used. The designers of Maple take the pragmatic view that it is better to have an actual program available for people to use, bugs and all, than to have a perfect program on the drawing board, or "ready any day now."

Every other computer algebra system also has bugs; often different ones, but remarkably many of these bugs are seen throughout all computer algebra systems, as a result of common design shortcomings. Probably the most useful advice I can give for dealing with this is *be paranoid*.[15] Check your results in at least two ways (the more the better). Don't just do the calculation again in the same way. Instead, do a simple case by hand, do selected cases numerically, substitute your solution back into the defining equation and look at the residual, do the same for the initial and boundary conditions, plot the results, and compare with physical experiment and the results of other people's calculations.

This will help you correct bugs in Maple, bugs in your own code, and bugs in your problem set-up. It may even help you correct bugs in your thoughts.

When you find a bug in Maple, report it by e-mail to
                    `support@maplesoft.com` .
If you can, isolate the bug in as small a code fragment as possible, and clearly explain what you think should be happening. The best bug reports include their own suggested fixes, and these get the highest priority.

Remark. If Maple is so "buggy," why do I recommend that people use it? And why do I use it so much myself (for twenty years, now)? Nearly all my research work uses Maple, sometimes only for trivial purposes but sometimes to make the central investigations of the paper. Obviously, Maple is very usable indeed, and sometimes indispensable. Indeed, despite the caution evident in this paragraph, Maple is one of the more reliable and well-tested large software packages. But people who uncritically believe what computers tell them deserve what they get: Garbage In, Gospel Out is a poor motto for our age. What we want instead is *Computer-Mediated Thinking* (CMT) for mathematics. In general, CMT means a human using a computer to help the human to think. My point is that skepticism is

---

[14]To be fair, `changevar` is intended to let the user do what she or he wants. Also, change-of-variables is often useful even if it isn't monotonic, surprisingly.

[15]This passage has not changed from the first edition, which is a good thing because it is quoted in [30, p. 239].

healthy in thinking. To help a human to think about mathematics, the most natural tools to use are computer algebra systems.

Exercises

1. Give a definition of a "bug in the program".

1.6 Evaluation Rules

Maple's evaluation rules, discussed fully in [44, Ch. 2], are designed to do what the user will expect in most situations. However, behaving as expected in *all* situations is impossible: Different people have different expectations in the same situations. The `eval` function and right quotation marks (') are used to customize evaluation. See `?spec_eval_rules` for a description of exceptions to the following.

1. Global variables are evaluated fully, at the top (interactive) level. For example, in an interactive session

    ```
    >   x := y;
    ```
    $$x := y$$
    ```
    >   y := z;
    ```
    $$y := z$$
    ```
    >   z := 3;
    ```
    $$z := 3$$
    ```
    >   x;
    ```
    $$3$$
    ```
    >   y;
    ```
    $$3$$

2. Right quotation marks prevent evaluation: A single evaluation merely removes the marks and does not go "all the way down."

    ```
    >   'x';
    ```
    $$x$$

    That is why the following construct allows you to "unassign" a name, or clear a variable.

    ```
    >   x := 'x';
    ```
    $$x := x$$
    ```
    >   x;
    ```
    $$x$$
    ```
    >   y;
    ```
    $$3$$

3. Local variables are evaluated *one* level in procedures. This is usually what is wanted. See Chapter 3 for a further discussion of global, local, and environment variables.

4. Arrays and tables have special evaluation rules. See ?eval for details.

5. Objects are *not* evaluated by the subs or subsop commands. For example,

   ```
   >   restart;
   >   f := sin(x);
   ```
   $$f := \sin(x)$$

   ```
   >   subs( x=0, f );
   ```
   $$\sin(0)$$

   Referring to that value (e.g., via %) will cause it to be fully evaluated; likewise, asking for an explicit evaluation with eval will do the job.

6. Sometimes, subs will do substitutions that are nonsensical:

   ```
   >   gp := diff( g(x), x );
   ```
   $$gp := \frac{\partial}{\partial x}\, g(x)$$

   ```
   >   subs( x=0, gp );
   ```
   $$\text{diff}(g(0),\, 0)$$

   ```
   >   %;
   ```

   ```
   Error, wrong number (or type) of parameters in function
   diff
   ```

   Maple provides another facility that is more careful, namely eval:

   ```
   >   gp0 := eval( gp, x=0 );
   ```
   $$gp0 := \left.\left(\frac{\partial}{\partial x}\, g(x)\right)\right|_{x=0}$$

   ```
   >   lprint( gp0 );
   ```

   ```
   eval(diff(g(x),x),{x = 0})
   ```

   ```
   >   eval( gp0, g=sin );
   ```

1

7. Evaluation can be prevented with right quotation marks (') and forced with eval as seen previously.

```
>  for i to 5 do i^2 end do;
```

$$1$$
$$4$$
$$9$$
$$16$$
$$25$$

```
>  i;
```

$$6$$

```
>  sum( i^3, i=1..n );
```

Error, (in sum) summation variable previously assigned,
second argument evaluates to 6 = 1 .. n

```
>  sum( 'i^3', 'i'=1..n );
```

$$\frac{1}{4}(n+1)^4 - \frac{1}{2}(n+1)^3 + \frac{1}{4}(n+1)^2$$

The above examples show that the eval command can be used to force the desired level of evaluation in exceptional circumstances. See also ?eval.

For special domains, such as matrices, hardware floating-point, Maple floating-point, complex numbers, or others, special evaluation routines are used, as follows.

The routine evalhf is used for hardware floating-point evaluation (for speed).

```
>  evalhf( sin(1) );
```
$$.841470984807896505$$

The routine evalf is used for Maple arbitrary-precision floating-point (for more accuracy than evalhf).

```
>  Digits := 20;
```
$$Digits := 20$$

```
>  evalf( sin(1) );
```
$$.84147098480789650665$$

Alternatively, you can pass the number of requested digits to evalf in an index to evalf.

```
>  evalf[40]( sin(1) );
```
$$.8414709848078965066525023216302989996226$$

Further, floats are *automatic* to some extent, so sin(1.) automatically evaluates to (about) 0.841. However, Pi/1.0 evaluates to 1.000000000π, not 3.141592654.

1.6.1   Working With Complex Numbers and Expressions

The routine `evalc` is used for complex expressions, but complex numbers are automatically converted to Cartesian form.

> `restart;`

> `(1+2*I)/(3+4*I);`

$$\frac{11}{25} + \frac{2}{25}\,I$$

> `arcsin( 2. );`

$$1.570796327 - 1.316957897\,I$$

> `ln( -1 );`

$$I\,\pi$$

The principal branches of the elementary functions, with closures on the branches chosen in a careful way, are used throughout Maple's numerics. See Appendix A.

> `x := a + b*I;`

$$x := a + I\,b$$

> `y := c + d*I;`

$$y := c + I\,d$$

> `pr := x*y;`

$$pr := (a + I\,b)\,(c + I\,d)$$

> `evalc( pr );`

$$a\,c - b\,d + I\,(a\,d + b\,c)$$

> `evalc( Re(pr) );`

$$a\,c - b\,d$$

> `evalc( Im(pr) );`

$$a\,d + b\,c$$

> `evalc( Re(1/x) );`

$$\frac{a}{a^2 + b^2}$$

and similarly for some others. Note that `evalc` assumes implicitly that names are *real-valued*, while most of the rest of Maple does not.

Why is '*I*' Maple's name for the square root of −1?

Maple is also a programming language, and since the designers of Maple are primarily programmers, their concerns are those of programmers. Since lower-case i is used extensively as an index in `for` loops, as is j, the Maple designers could not bear to use either for the imaginary unit. They had no such need for uppercase I, and some also have an aversion to using two-character codes such as _i. What is needed is a mathematically typeset math-italic i, but that isn't available in ASCII.

> The choice of I instead of i for the imaginary unit concedes to the widespread use of the identifier i for other purposes.

—JTC1/SC22/WG14 - C Draft Rationale N897 for C99

The preempting of I for the imaginary unit causes some difficulty if you wish to use I as a name for an integral, or as an identity matrix or operator, for example. However, you *can* do it, since I is now predefined in Maple in a way that can be changed. If you want to use the variable I for your own purposes, you could issue the command interface(imaginaryunit=j) first, which frees I for your own use (and locks up j, but you can use whatever you like). In the following, suppose that A is a predefined 2-by-2 matrix. Then we can use I as the identity matrix as follows.

```
>   interface( imaginaryunit=j );
>   with( LinearAlgebra ):
>   I := Matrix( 2, 2, shape=identity );
```

$$I := \begin{bmatrix} 1 & 0 \\ 0 & 1 \end{bmatrix}$$

```
>   3*I;
```

$$\begin{bmatrix} 3 & 0 \\ 0 & 3 \end{bmatrix}$$

1.6.2   Inert Functions

Sometimes you don't want Maple to evaluate an expression at all, or at least until you tell Maple to do so later. A useful mechanism for doing this is provided by the concept of an "inert function." Typically, the name of an inert function in Maple is the same as that of the active function it represents, except that the initial letter is capitalized: Int, Sum, Diff, Svd, and Eigenvals are examples. This nomenclature rule is not universal.

A typical use of an inert function call is to prevent symbolic evaluation of the problem and allow later numerical procedures to be invoked. This can save time spent symbolically processing the input in the case where such processing would necessarily fail. Consider the following example, where we first try to integrate a function that has no closed-form antiderivative. We time the results using time(). The following is the standard syntax for the use of this routine

```
>    restart;
>    Digits := 30;
```

$$Digits := 30$$

```
>    ratio := Vector( 10 ):
>    for k to 10 do
>        n := rand()/10^13;
>        m := rand()/10^13;
>        tn := time( evalf(int( t*tan(n*t), t=0..1)) );
>        tm := time( evalf(Int( t*tan(m*t), t=0..1)) );
>        ratio[k] := evalf( tn/tm, 3 );
>    end do:
>    seq( ratio[k], k=1..10 );
```

$$3.29, \ 1.95, \ 1.93, \ 1.88, \ 1.87, \ 1.94, \ 1.82, \ 1.91, \ 1.82, \ 1.95$$

The first time is larger than the others because it includes time spent loading the routines, etc. So, in the active (not inert) case, about half the time was spent on symbolic processing. Of course, if it had succeeded, then usually the resulting numerical evaluation would be more efficient.

Let us consider a case where we expect symbolic processing to fail: Computation of the eigenvalues of a random five-by-five matrix.

```
>    restart;
>    with( LinearAlgebra ):
>    A := RandomMatrix( 5, 5 );
```

$$A := \begin{bmatrix} -66 & -65 & 20 & -90 & 30 \\ 55 & 5 & -7 & -21 & 62 \\ 68 & 66 & 16 & -56 & -79 \\ 26 & -36 & -34 & -8 & -71 \\ 13 & -41 & -62 & -50 & 28 \end{bmatrix}$$

We should not bother to try to find the exact eigenvalues of this, because they will just be expressed as a RootOf containing the characteristic polynomial. It is well known that finding the roots of polynomials is much harder than finding the eigenvalues of the matrix directly [58].

```
>    Eigenvalues( A );
```

$$\begin{bmatrix} \text{RootOf}(\%1, \ index = 1) \\ \text{RootOf}(\%1, \ index = 2) \\ \text{RootOf}(\%1, \ index = 3) \\ \text{RootOf}(\%1, \ index = 4) \\ \text{RootOf}(\%1, \ index = 5) \end{bmatrix}$$

$$\%1 := 4820471082 + 31455750 \ \_Z - 290950 \ \_Z^2 - 6369 \ \_Z^3 + 25 \ \_Z^4 + \_Z^5$$

Note the use of the label %1 to prevent duplicate printing of a large expression. Here the expression is just the characteristic polynomial, which we expect to be irreducible; moreover, finding numerical approximations to roots of polynomials can be expensive because they are often ill-conditioned. But finding approximations to the eigenvalues of a numerical matrix is easy, and usually numerically stable:

```
>   Eigenvalues( evalf(A) );
```

$$\begin{bmatrix} -90.0602177745938093 + 0.\ I \\ -47.5998797786374724 + 64.7235972142707540\ I \\ -47.5998797786374724 - 64.7235972142707540\ I \\ 80.1299886659341780 + 43.2589992370597898\ I \\ 80.1299886659341780 - 43.2589992370597898\ I \end{bmatrix}$$

Similarly, singular values (see [24]) may be computed numerically:

```
>   SingularValues( evalf(A) );
```

$$\begin{bmatrix} 163.700863169737346 \\ 124.040336672716720 \\ 101.660724732306392 \\ 89.3722828898417135 \\ 26.1288035233601441 \end{bmatrix}$$

Similar considerations hold for evaluating sums as for evaluating integrals. If we call sum, we are asking for closed-form antidifferences (see Section 2.4), which can take a long time to fail. If what we want is a numerical sum, it is best simply to do as follows. [Note that we issued a `restart` earlier, so here k is unassigned and thus we can use it in Sum without quoting it to prevent evaluation.]

```
>   S := Sum( 1/k^2, k=1..infinity );
```

$$S := \sum_{k=1}^{\infty} \frac{1}{k^2}$$

```
>   evalf( S );
```

$$1.644934067$$

Of course, Maple knows that the sum is $\pi^2/6$, and symbolic processing would succeed here. Consider now the following counterintuitive result.

```
>   S := Sum( 1/sqrt(k), k=1..infinity );
```

$$S := \sum_{k=1}^{\infty} \frac{1}{\sqrt{k}}$$

```
>   value( S );
```

$$\infty$$

The command `value` converts an inert function to its active form, forcing symbolic processing. Here Maple correctly determines that the series diverges to ∞. What does `evalf(Sum(...))` do?

```
>  evalf( S );
```
$$-1.460354509$$

It returns a negative answer! This is because the sequence acceleration method used (Levin's u-transform [57]) will sometimes give a numerical value to a formally divergent series. This is often useful, because formally divergent sums may (with appropriate convergence acceleration techniques) sum to physically interesting values. Here, adding positive terms to get a negative value does not seem to make sense—but this can be justified via analytic continuation (of the Riemann ζ-function—evaluate $\zeta(\frac{1}{2})$ in Maple to see how close the above answer is). It is clear that you must use `evalf(Sum)` with caution. See Section 2.4 for a program to do summation of divergent series another way.

Here is another example of inert functions, showing conversion to the active form using `value`.

```
>  B := Int( ln(x)/(1-x^2), x );
```
$$\int \frac{\ln(x)}{1-x^2}\, dx$$

```
>  value( B );
```
$$\frac{1}{2}\,\mathrm{dilog}(x) + \frac{1}{2}\,\mathrm{dilog}(x+1) + \frac{1}{2}\ln(x)\ln(x+1)$$

Finally, here is an example showing inert functions used in the evaluation of functions over a finite field.

```
>  p := x^5 + x^3 + x;
```
$$p := x^5 + x^3 + x$$

```
>  Factor(p) mod 2;
```
$$x\,(x^2 + x + 1)^2$$

The `Factor` command was inert, but the postfix `mod` operator evaluated it over the integers mod 2.

```
>  q := diff(p,x) mod 2;
```
$$q := x^4 + x^2 + 1$$

```
>  Gcd(p, q);
```
$$\mathrm{Gcd}(x^5 + x^3 + x,\ x^4 + x^2 + 1)$$

```
>  Gcd(p,q) mod 2;
```
$$x^4 + x^2 + 1$$

Exercises

1. If $z = x + iy$, where x and y are real numbers, use Maple to find $\Re[(z + 1)/(2 + z^2)]$. (Note: $\Re(z)$ is the real part of z. Likewise, $\Im(z)$ is the imaginary part. The Maple commands are Re and Im.)

2. If A is the matrix below, find $I + A + A^2 + A^3$. Note that it is obvious in this context that I is meant to be the identity matrix, not the square root of -1.

$$A = \begin{bmatrix} 1 & 1 & a \\ 1 & a & 0 \\ a & 0 & 1 \end{bmatrix}$$

3. Verify that the eigenvalues and singular values of the 5-by-5 random matrix given in Section 1.6.2 are correct.

4. Define a variable f to be the expression $x + \sin(2x)$. Use subs to evaluate this expression if $x = 0, 1$, and $-\pi$. Then use eval with a second argument to do the same thing.

5. Compare the execution times of evalf(int(1/(1+t^64), t=0..1) and the inert form. Explain.

6. Compare the execution times of evalf(sum(exp(-k), k=1..infinity)) with its inert form. Also examine add(exp(-1.0*k), k=1..4000) and the for-loop
   ```
   >   s:=0: for k to 4000 do s:=s+exp(-1.0*k):
   >   end do:
   ```
   Be careful to use a colon (:) and not a semicolon (;) at the end of the end do:, because otherwise all the intermediate results will be printed, which will disrupt the timings.

7. Compute the singular values of a random 100 by 100 matrix with SingularValues(evalf(A)). Compare the execution speed with MATLAB if you have access to it. MATLAB is designed to be very fast at this type of computation. Maple now uses a link to the NAG library.

1.7   The assume Facility

> 'I can be of little avail to your lordship if you give me unsufficient premises to reason from. But worse than tell it not to me, I fear you tell it not truly to yourself.'
> —E. R. Eddison, *A Fish Dinner in Memison*, Chapter XII.

Maple is a computer *algebra* language. It is now apparent that many people try to use Maple (or other computer algebra languages) for *analysis* [14]. This has caused some difficulty in the past, since there are differences between what is considered correct if you are doing algebra and what is considered correct if you are doing analysis. For example, consider the solution of $(k^4 + k^2 + 1)x = k^4 + k^2 + 1$. If k is real, then we can say unequivocally that $x = 1$, but if k might be one of the complex roots of $k^4 + k^2 + 1$, we must add a proviso to the result $x = 1$ for complete correctness, namely that $x = 1$, provided that $k^4 + k^2 + 1 \neq 0$. This is an example of a correct analytical result. However, the *algebraic* approach to this problem would be to consider k as an *indeterminate*. Thus k has *no* value, and so it is perfectly legal to divide by this polynomial in k, so long as it isn't the zero polynomial (which it isn't, since not all the coefficients are zero). In that case, $x = 1$ with no provisos at all. Computer algebra systems *often take this point of view*, and it is only recently that the analytical viewpoint has been considered in many cases.

A related problem is the evaluation of integrals and sums that depend on parameters. It is possible that—for some values of the parameters—the integral is finite, and—for others—that the integral does not converge. For example, consider

$$\int_0^\infty \exp(-st)\, dt \ .$$

In a naive attempt to solve this with algebra, we find the antiderivative for $\exp(-st)$, which is $-\exp(-st)/s$, and plug in the limits of the integration: $-1/s$ at the bottom and $\ldots$, well, something at the upper limit: $-\exp(-s\infty)$. It is easy to overlook the fact that $\Re(s)$ might be nonpositive and indeed there has not been, until recently, any way of telling Maple about this analytic, or geometric, information.

We need some way of telling Maple about our assumptions on s, in this case that (for example) $s > 0$. Once Maple knows about the properties of the variables, it can proceed *correctly*. Prior to Maple V Release 2, Maple would simply give you the answer $1/s$ for this integral, which would be incorrect if $s \leq 0$. Since Maple V Release 2, Maple makes some attempt to inform you that it needs to know more about s.

```
>   restart;

>   int( exp(-s*t), t=0..infinity );
```

```
Definite integration: Can't determine if the integral
is convergent. Need to know the sign of --> s. Will
now try indefinite integration and then take limits.
```

$$\lim_{t \to \infty} -\frac{e^{(-s\,t)} - 1}{s}$$

and the user now knows that Maple can do *something* with the integral but it needs to know more about s before it can evaluate the limit.

Anthea laughed. 'Timourous scrupulosities! 'Twas meant, if it were not said.'

—E. R. Eddison, *A Fish Dinner in Memison*, Chapter XI

This type of behaviour is not universal: For example, consider

```
>   int( exp(-t)*t^(x-1), t=0..infinity );
```

```
Definite integration: Can't determine if the integral
is  convergent. Need to know the sign of --> x. Will
now try indefinite integration and then take limits.
```

$$\Gamma(x)$$

Sometimes it succeeds, because of implicit assumptions in `limit`.

```
>   int( exp(-s*t)/(1+t), t=0..infinity );
```

$$\lim_{t\to\infty} -e^s \operatorname{Ei}(1, s+st) + \operatorname{Ei}(1, s)e^s$$

Sometimes it doesn't. Note that there were no warning messages that time; it is up to the user to infer that something more can be done if Maple is told about the sign of s.

These problems are representative of the problems the `assume` facility, and, new to Maple 7, the `assuming` command, are meant to address. The following session illustrates some of the simple things you can do with `assume`.

```
>   assume( s>0 );
```

As we would expect, Maple deduces that s is real from our assumption that it is positive.

```
>   Int( exp(-s*t)/(1+t), t=0..infinity )
>       = int( exp(-s*t)/(1+t), t=0..infinity ) ;
```

$$\int_0^\infty \frac{e^{(-\tilde s\, t)}}{1+t}\, dt = e^{\tilde s}\operatorname{Ei}(1,\tilde s)$$

The tilde (or twiddle) ~ after the s here is Maple's way of reminding you that you have made assumptions about s.

The example below shows that Maple can figure out that $s+1 > 0$ if it knows that $s > 0$.

```
>   Int( exp(-s*(s+1)*t)/(1+t), t=0..infinity )
>   = int( exp(-s*(s+1)*t)/(1+t), t=0..infinity );
```

$$\int_0^\infty \frac{e^{(-\tilde s\,(\tilde s+1)\,t)}}{1+t}\, dt = e^{(\tilde s^2)}e^{\tilde s}\operatorname{Ei}\left(1,\tilde s^2+\tilde s\right)$$

At this point, we see why I don't like twiddles. It is too easy to read $s\tilde{}2$ as s^{-2}. A cure is mentioned below. Maple knows that $s > 0$, but we have not told it that $s > 1$, as we see in the following:

```
>    Int( exp(-s*(s-1)*t)/(1+t), t=0..infinity )
>    = int( exp(-s*(s-1)*t)/(1+t), t=0..infinity );
```

Definite integration: Can't determine if the integral is
convergent. Need to know the sign of --> s*(s-1). Will
now try indefinite integration and then take limits.

$$\int_0^\infty \frac{e^{(-s\tilde{\,}(s\tilde{\,}-1)\,t)}}{1+t}\,dt =$$

$$\lim_{t\to\infty} -e^{(s\tilde{\,}^2-s\tilde{\,})}\,\mathrm{Ei}(1,\,s\tilde{\,}^2-s\tilde{\,}+s\tilde{\,}^2 t-s\tilde{\,}t) + \mathrm{Ei}(1,\,s\tilde{\,}^2-s\tilde{\,})\,e^{(s\tilde{\,}(s\tilde{\,}-1))}$$

The above shows that Maple's built-in routines are capable, to some extent, of
using assumed knowledge. You may wish to write your own programs to test
knowledge of parameters: the routines to use are is and isgiven. See the help
file for assume for details.

> 'No,' she said, looking upon them daintily: 'they have too many
> twiddles in them: like my Lord Lessingham's distich.'
> —E. R. Eddison, *Mistress of Mistresses*, Chapter VII.

Luckily, under the Options/Assumed Variables menu, we can choose "phrase"
instead of twiddles. You can also do this from the command line by issuing
the command interface(showassumed = 2); and this can be put in a file
called maple.ini (or, on Unix, .mapleinit). That file is executed every time
you start Maple if the file can be found in the path. One way to ensure this in
Windows is to put the file maple.ini in the Maple bin directory.

Choosing this option for displaying assumed variables gives rise to a neater
and more readable answer:

```
>    assume( s > 1 );
>    sum( k^(-s), k=1..infinity );
```

$$\zeta(s)$$

with assumptions on s

Exercises

1. Assume $x > 0$, and verify that Maple knows then how to evaluate
   $\int_0^\infty \exp(-t)t^{x-1}\,dt$.

2. Assume that $s > 2$. Use Maple to compute the signum of s.

3. If x and y satisfy $x < -|y|$ and y is real, is the matrix

$$A = \begin{bmatrix} x & y \\ y & x \end{bmatrix}$$

negative semidefinite? See ?IsDefinite. The answer to this exercise can
be done in one line, using assuming.

Useful One-Word Commands

Maple has many built-in "black boxes" for the simplification of algebraic expressions, solution of algebraic and calculus problems, standard operations from calculus, manipulations from the calculus of finite differences, solution of linear algebra problems, and evaluation of functions. This chapter examines some of the most useful of these black boxes.

There are groups of commands that perform similar (but not identical) tasks. These are collected together in sections here, so that their similarities and differences can be highlighted. Sometimes the differences in speed or quality of solution from one routine to another can be dramatic. One of the hardest things to learn in Maple is when to use a particular routine over another. The purpose of this chapter is to help you to gain experience in making such choices.

2.1 Simplification

> 'That question,' said Vandermast, 'raiseth problems of high dubitation: a problem *de natura substantiarum*; a problem of selfness. Lieth not in man to resolve it, save so far as to peradventures, and by guess-work.'
>
> —E. R. Eddison, *A Fish Dinner in Memison*, Chapter VIII.

Simplification is, in general, an intractable problem. It has been proved that it is impossible to write a computer program that recognizes when arbitrary input expressions are equivalent to zero [22]. This result, which is concerned with an infinite class of input expressions, translates into real difficulty in dealing with specific input expressions. For example, is $\log \tan(x/2 + \pi/4) - \operatorname{arcsinh} \tan x = 0$?

[It is, if $-\pi/2 < x < \pi/2$, but not if $\pi/2 < x < 3\pi/2$, for example.] Certainly recognizing zero when you see it seems fundamental to simplification.

2.1.1 `normal`

If we restrict the class of functions we deal with, we can provide a *normal form* of representation. Zero is represented uniquely in a normal form: There are no nontrivial representations of zero in a normal form. For still further restricted classes of functions[1] we can provide a *canonical form* of representation. Each function is represented uniquely in a canonical form. Polynomials over the rationals have several possible canonical forms. For example, collect all like terms and sort them in ascending order. The Maple command `normal` puts multivariate rational polynomials with integer coefficients in a normal form, cancelling common integer-coefficient factors by use of GCDs [22]. This command should be used throughout computations with rational polynomials, since this keeps the size of the intermediate expressions down (generally speaking). For example,

```
>  p := expand( (x+1)^3*(x+2)^2*(x+3) );
```
$$p := x^6 + 10\,x^5 + 40\,x^4 + 82\,x^3 + 91\,x^2 + 52\,x + 12$$
```
>  q := diff( p, x );
```
$$q := 6\,x^5 + 50\,x^4 + 160\,x^3 + 246\,x^2 + 182\,x + 52$$
```
>  r := p/q;
```
$$r := \frac{x^6 + 10\,x^5 + 40\,x^4 + 82\,x^3 + 91\,x^2 + 52\,x + 12}{6\,x^5 + 50\,x^4 + 160\,x^3 + 246\,x^2 + 182\,x + 52}$$
```
>  normal( r );
```
$$\frac{1}{2}\,\frac{x^3 + 6\,x^2 + 11\,x + 6}{3\,x^2 + 13\,x + 13}$$

Note that `normal(ab)` could simplify the result to either $a+b$ or $b+a$. In a normal form you cannot rely on more than zero recognition: Without further work, in general, a normal form is not a canonical form. The following larger example shows this.

```
>  Int( 1/(2+sqrt(x)), x ) = int( 1/(2+sqrt(x)), x ) + C;
```
$$\int \frac{1}{2+\sqrt{x}}\,dx = -2\ln(-4+x) + 2\sqrt{x} + 2\ln(-2+\sqrt{x}) - 2\ln(2+\sqrt{x}) + C$$

One way to test to see whether Maple got the correct answer is to differentiate both sides mechanically and see whether they are the same. This is a *necessary*

---

[1] Well, not technically. Arthur Norman informs me that if you have a normal form on a countable set (such as rational functions over the integers), then we can use the "British Museum algorithm": Take the object to be canonicalized, and then, in the counting order, compare it to each of the class of objects in turn, putting the difference into a normal form. Declare the first object in the list that gives a zero normal form of the difference to be the canonical representation of the given object.

condition, but not a sufficient condition: It is possible that Maple (or any other computer algebra system) will produce an answer that will pass this test but still be wrong. See Section 2.3.2 for further discussion.

```
>  diff( %, x );
```

$$\frac{1}{2+\sqrt{x}} = -2\,\frac{1}{-4+x} + \frac{1}{\sqrt{x}} + \frac{1}{\sqrt{x}\,(-2+\sqrt{x})} - \frac{1}{\sqrt{x}\,(2+\sqrt{x})}$$

Those expressions do *not* look equal. However, we can try to see whether each simplifies to the same thing, as follows.

```
>  normal( % );
```

$$\frac{1}{2+\sqrt{x}} = -\frac{-8+2\,x+4\,\sqrt{x}-x^{(3/2)}}{(-4+x)\,(-2+\sqrt{x})\,(2+\sqrt{x})}$$

Now, with a little work, a human can show that both sides are equal, but the `normal` command did not simplify both expressions to the same form (i.e., `normal` does not produce a canonical form for expressions). It will, however, simplify the difference between these two expressions to zero.

```
>  normal( (lhs - rhs)( % ) );
```

$$0$$

Exercises

1. Create a worksheet with only two commands in it: `restart;` and `normal( 1/(a-b) );`. Execute the pair of commands (restart then normal) over and over, until you see the result change from one time to the next. This is because `normal` chooses its "main variable" *randomly* and tries to make the equation monic in that main variable. This causes ordering problems in worksheets.

2. Plot Maple's answer to $\int 1/(1+\sqrt{x})\,dx$, and show that it is real for $x > 0$ because the imaginary parts of the logarithms cancel.

3. Try to find other examples where `normal` will give two different representations for the same expression, but when called on the difference between the two expressions will give zero. [The derivative of the integral of $1/(t^6-1)$, from the second sample Maple session in Chapter 1, did this, if you didn't use the "expanded" option.]

2.1.2 collect

The Maple command `collect` can be used to put multivariate polynomials into a canonical form, with all coefficients of similar terms collected. This function (or

expand) can be called to improve the presentation of a polynomial. The routines
coeff or coeffs can be used to pick off the coefficients.

The most useful feature of collect is that you can apply any function you
like to each coefficient as the polynomial is collected.

```
>   p := 1 + x + 3 + 5*x + 6*y + 17*y^2 + 35*x
>           + 52*x^2 + 99*x*y + (x+y)^3 ;
```

$$p := 4 + 41\,x + 6\,y + 17\,y^2 + 52\,x^2 + 99\,x\,y + (x+y)^3$$

```
>   collect( p, x );
```

$$x^3 + (52 + 3\,y)\,x^2 + (41 + 99\,y + 3\,y^2)\,x + 4 + 6\,y + 17\,y^2 + y^3$$

```
>   collect( p, y );
```

$$y^3 + (17 + 3\,x)\,y^2 + (3\,x^2 + 99\,x + 6)\,y + 4 + 41\,x + x^3 + 52\,x^2$$

```
>   collect( p, [x,y] );
```

$$x^3 + (52 + 3\,y)\,x^2 + (41 + 99\,y + 3\,y^2)\,x + 4 + 6\,y + 17\,y^2 + y^3$$

```
>   collect( p, [x,y], distributed );
```

$$x^3 + 52\,x^2 + 3\,y\,x^2 + 41\,x + 99\,x\,y + 3\,y^2\,x + 4 + 6\,y + 17\,y^2 + y^3$$

Applying functions to coefficients.

The command collect(p,x,<function>) applies the function <function> to
each coefficient, once collected. Sensible functions to use in this context include
factor and simplify.

An example from analysis of numerical methods for IVP. Consider the logistic
differential equation again, but now imagine that we are trying to analyze the
behaviour of Euler's method $x_{n+1} = x_n + hx'_n$ rather than trying to solve this
equation exactly.

```
>   restart;
```

```
>   Logistic := x -> diff(x,t) - x*(1-x);
```

$$Logistic := x \rightarrow (\tfrac{\partial}{\partial t}\,x) - x\,(1 - x)$$

Think of v as x_n, and w as x_{n+1}. We use alias here to try to compress the
output. If we simply assigned $w := v + hv(1 - v)$, then when we typed w Maple
would interpret it as $v + hv(1 - v)$, but the advantage of alias is that in addition,
when Maple recognizes that it has come up with $v + hv(1 - v)$ it will display it
as w.

```
>   alias( w = v + h*v*(1-v) );
```

$$w$$

The following is the cubic Hermite interpolant between v and w.

```
>   interpolant := v + t*v*(1-v)
>        + (v*(1-v) - w*(1-w))/h^2*t^2*(h-t);
```

$$interpolant := v + t\,v\,(1-v)$$
$$+ \frac{(v\,(1-v) - w\,(1-v-h\,v\,(1-v)))\,t^2\,(h-t)}{h^2}$$

We verify that this interpolant matches the solution and its derivative at both ends:

```
>   eval( interpolant, t=0 );
```
$$v$$

```
>   eval( interpolant, t=h );
```
$$w$$

```
>   eval( diff( interpolant, t ), t=0);
```
$$v\,(1-v)$$

```
>   eval( diff( interpolant, t ), t=h );
```
$$w\,(1-v-h\,v\,(1-v))$$

Notice that $\texttt{alias}$ did not detect that $-v - hv(1-v)$ is $-w$; but we can see that, and know thereby that the interpolant matches the function and the derivative at both ends.

The *residual*, or *defect*, tells us how much the interpolated numerical solution fails to satisfy the given equation.

```
>   defect := Logistic( interpolant );
```

$$defect := v\,(1-v) + \frac{2\,\%1\,t\,(h-t)}{h^2} - \frac{\%1\,t^2}{h^2} -$$
$$(v + t\,v\,(1-v) + \frac{\%1\,t^2\,(h-t)}{h^2})$$
$$(1 - v - t\,v\,(1-v) - \frac{\%1\,t^2\,(h-t)}{h^2})$$
$$\%1 := v\,(1-v) - w\,(1-v-h\,v\,(1-v))$$

Note that Maple has detected (as happened once in Chapter 1) a large common subexpression in this result, and has chosen to label it %1 and print it only once to save space and to help the reader understand the whole expression.

The interpolant is useful only if $t_n < t < t_n + h$, so we put $\theta = (t - t_n)/h$ and impose $0 \le \theta \le 1$. We collect in powers of h, and because it is simpler to understand each coefficient if it is in factored form, we apply the function $\texttt{factor}$ to each coefficient:

```
>   simpler := collect( eval(defect,t=theta*h), h, factor );
```

$$simpler := v^4 \, (v-1)^4 \, \theta^4 \, (-1+\theta)^2 \, h^6$$
$$- 2 \, v^3 \, (2\,v-1) \, (v-1)^3 \, \theta^4 \, (-1+\theta)^2 \, h^5 + v^2 \, \theta^3 \, (-1+\theta)$$
$$(v-1)^2$$
$$(4\,v^2\,\theta^2 - 4\,v^2\,\theta + 2\,v^2 - 4\,\theta^2\,v + 4\,v\,\theta - 2\,v + \theta^2 - \theta)\,h^4$$
$$- v^2\,\theta^2\,(2\,\theta+1)\,(-1+\theta)\,(2\,v-1)\,(v-1)^2\,h^3$$
$$+ v\,\theta\,(-1+\theta)\,(v-1)\,(4\,v^2\,\theta - 2\,v^2 - 4\,v\,\theta + 2\,v + \theta)\,h^2$$
$$+ 3\,v\,\theta\,(2\,v-1)\,(v-1)\,(-1+\theta)\,h$$

It is quite easy to interpret the result above, but numerically only the first few powers of h are important:

```
>   series( simpler, h, 3 );
```

$$3\,v\,\theta\,(2\,v-1)\,(v-1)\,(-1+\theta)\,h$$
$$+ v\,\theta\,(-1+\theta)\,(v-1)\,(4\,v^2\,\theta - 2\,v^2 - 4\,v\,\theta + 2\,v + \theta)\,h^2$$
$$+ O(h^3)$$

From here it is easy to see that the defect is zero if $v = 0$ or $v = 1$ (steady states of the logistic equation), and also at $\theta = 0$ and $\theta = 1$ (the ends of the numerical step), and that asymptotically the maximum defect will occur when $\theta = 1 - \theta$ or $\theta = 1/2$. [Aside for those who are not numerical analysts: Because the residual is $O(h)$ as $h \to 0$, we say that Euler's method is a first-order method. The Gröbner–Alexeev[2] nonlinear variation of constants formula (see [26]) shows that if the defect is $O(h)$ on a compact interval, then the global error will also be $O(h)$.]

Using collect to hide complicated coefficients. The following more complicated example is discussed more fully in [18], and modules are discussed in Chapter 3, but for now just observe the use of the applied function in collect to replace unwieldy expressions with more understandable labels. The routine in Figure 2.1, which I have tentatively called LEM (for Large Expression Management), will generate a module that, when its exports are bound via with, or used in long form as below, will provide a routine named veil, the auxiliary routine unveil, and an index lastUsed pointing at the last computed entry in the sequence. There is quite a bit to say about the Maple elements used in this small object-oriented program. See Section 3.9.3. Maple is not an object-oriented language per se, but it does provide support for programming in that manner. The purpose of the routine veil is to replace its argument with an unassigned label (the name of the label to use is passed to LEM, and hence you can have more than one sequence of labels in the same session) and to remember the value hidden under the label in the asso-

---

[2]Yes, the same Gröbner as in Gröbner bases.

```
macro('NOTPINS'=LEM): 'NOTPINS' := proc( C::name )
   module()
      export veil, unveil, lastUsed;
      local auxiliary, labelledValues, str;
      if assigned( C ) then
         # Use a local here just so the line isn't too long.
         str := cat(sprintf("label %a is assigned a value already.\n", C),
            sprintf("Save its contents and unassign( %a );\n", C),
            sprintf("There is no need to repeat the call to %a.",'NOTPINS'));
         WARNING( str );
      end if;
      lastUsed := 0; # Begin with nothing recorded
      labelledValues := table():

      unveil := proc( c, ilevel::{nonnegint,infinity} )
         local a, b, i, level;
         description "reveal expressions hidden behind labels.";
         level := 'if'( nargs<2, 1, min(lastUsed+1,ilevel) );
         a := c;
         # Always do at least 1
         b := eval( a, [seq(C[i]=labelledValues[i],i=1..lastUsed)] );
         for i from 2 to level while not Testzero(a-b) do
            a := b;
            b := eval( a, [seq(C[i]=labelledValues[i],i=1..lastUsed)] );
         end do;
         return b;
      end proc;

      veil := proc( coefficient )
         local i, s, c;
         description "hide expressions behind labels.";
         # Recognize zero if we can, so that we don't hide zeros.
         c := Normalizer( coefficient );
         # Remove the integer content and sign so that we don't hide them.
         i := icontent( c );
         s := sign( c );
         # Hide it only if it's not just a constant
         if s*i=c then return c end if;
         s*i*auxiliary( s*c/i );
      end proc;

      # Scope lastUsed etc and use option remember to detect duplicates.
      auxiliary := proc( c )
         option remember;
         labelledValues[ lastUsed ] := c;
         lastUsed := lastUsed + 1;
         C[ lastUsed ]
      end proc:
   end module:
end proc:  macro('NOTPINS'='NOTPINS'):
```

Figure 2.1: A Maple program that facilitates replacement of unwieldy expressions with simple labels

ciated computation sequence. You can query the computation sequence by using the routine `unveil`.

We can then use `collect`, together with `veil`, to label unwieldy expressions for convenience of use or understanding, as exemplified below. Be aware that Maple may choose to order things differently in your session, if you choose to execute these commands. This, then, means that the coefficients are veiled in a different order and thus the sequence of constants may differ from session to session.

```
> restart;
```

The `currentdir` command sets the path.

```
> currentdir("D:/books/ess/programs");
> read "veil.mpl";
```

Construct a procedure to hide expressions under the labels $K_1, K_2, \ldots$, but first as a test of the error-checking, assign something to K (as may happen by accident):

```
> K := table();
```

$$K := \text{table}([])$$

```
> VK := LEM( K ):
```

```
Warning, label K is assigned a value already.
Save its contents and unassign( K );
There is no need to repeat the call to LEM.
```

```
> unassign( K ):
```

Now construct another procedure, to be used at the same time (just to show that we can), to label expressions with a different constant, C:

```
> VC := LEM( C ):
```

The following complicated expression is to be simplified by viewing it as a polynomial in x and y only:

```
> p1 := randpoly( [x,y,z], dense, degree=7 );
```

$$\begin{aligned}
p1 := {} & 81 + 28\,x - 55\,x^6\,y - 37\,x^6\,z + 40\,y\,z + 97\,x^5\,y^2 + 79\,x^5\,y \\
& + 56\,x^5\,z^2 + 49\,x^5\,z + 25\,y^2\,z + 9\,y\,z^2 + 57\,x^4\,y^3 + 61\,y \\
& + 30\,z - 85\,x^7 - 35\,x^6 + 4\,y^2 + z^2 + 63\,x^5 + 22\,y^3 + 88\,z^3 \\
& + 50\,x^5\,y\,z - 59\,x^4\,y^2\,z - 8\,x^4\,y\,z^2 - 93\,x^4\,y\,z - 5\,x^3\,y^3\,z \\
& - 61\,x^3\,y^2\,z^2 - 50\,x^3\,y^2\,z - 18\,x^3\,y\,z^3 + 31\,x^3\,y\,z^2 - 26\,x^3\,y\,z \\
& + 66\,x^4 + 68\,y^4 + 11\,z^4 - 61\,x^3 - 32\,y^5 + 45\,x^4\,y^2 + 92\,x^4\,y \\
& + 43\,x^4\,z^3 - 62\,x^4\,z^2 + 77\,x^4\,z + 40\,y^3\,z - 73\,y^2\,z^2 + 25\,y\,z^3 \\
& + 54\,x^3\,y^4 + 99\,x^3\,y^3 - 12\,x^3\,y^2 - 62\,x^3\,y + x^3\,z^4 - 47\,x^3\,z^3 \\
& - 91\,x^3\,z^2 - 47\,x^3\,z + 94\,y^4\,z - 36\,y^3\,z^2 - 43\,y^2\,z^3 - 55\,y\,z^4
\end{aligned}$$

$$+ 41\,x^2\,y^5 - 90\,x^2\,y^4 + 94\,x^2\,y^3 - 84\,x^2\,y^2 + 85\,x^2\,y$$
$$+ 49\,x^2\,z^5 + 78\,x^2\,z^4 + 17\,x^2\,z^3 + 72\,x^2\,z^2 - 99\,x^2\,z - 66\,y^5\,z$$
$$+ 39\,y^4\,z^2 - 98\,y^3\,z^3 - 88\,y^2\,z^4 + 62\,y\,z^5 - 86\,x\,y^6 + 80\,x\,y^5$$
$$- 29\,x\,y^4 - 47\,x\,y^3 + 43\,x\,y^2 - 58\,x^2\,y^4\,z + 53\,x^2\,y^3\,z^2$$
$$- x^2\,y^3\,z + 83\,x^2\,y^2\,z^3 - 86\,x^2\,y^2\,z^2 + 23\,x^2\,y^2\,z + 19\,x^2\,y\,z^4$$
$$- 50\,x^2\,y\,z^3 + 88\,x^2\,y\,z^2 - 53\,x^2\,y\,z + 30\,x\,y^5\,z + 72\,x\,y^4\,z^2$$
$$+ 66\,x\,y^4\,z - 91\,x\,y^3\,z^3 - 53\,x\,y^3\,z^2 - 19\,x\,y^3\,z + 68\,x\,y^2\,z^4$$
$$- 72\,x\,y^2\,z^3 - 87\,x\,y^2\,z^2 + 79\,x\,y^2\,z - 66\,x\,y\,z^5 - 53\,x\,y\,z^4$$
$$- 61\,x\,y\,z^3 - 23\,x\,y\,z^2 - 37\,x\,y\,z + 62\,z^5 - 85\,x^2 + 9\,y^6$$
$$- 78\,z^6 - 61\,y^7 + 40\,z^7 + 31\,x\,y - 34\,x\,z^6 - 42\,x\,z^5 + 88\,x\,z^4$$
$$- 76\,x\,z^3 - 65\,x\,z^2 + 25\,x\,z - 60\,y^6\,z + 29\,y^5\,z^2 + 78\,y^4\,z^3$$
$$- 17\,y^3\,z^4 + 5\,y^2\,z^5 - 59\,y\,z^6$$

We choose to label its coefficients using K. We do this by applying the function veil to each coefficient of the polynomial, considered as a polynomial in x and y, by using collect. We refer to the K-version of the routine veil by its long name, VK:-veil, as follows:

```
>    compact1 := collect( p1, [x,y], distributed, VK:-veil );
```

$$compact1 := -55\,x^6\,y + 97\,x^5\,y^2 + 57\,x^4\,y^3 - 85\,x^7 + 54\,x^3\,y^4$$
$$+ 41\,x^2\,y^5 - 86\,x\,y^6 - 61\,y^7 - K_2\,x - K_4\,y - K_5\,x^6 + 7\,K_7\,x^5$$
$$- K_8\,y^3 - 3\,K_{27}\,y^6 + K_1 + K_6\,y^2 + K_3\,x^5\,y + K_{19}\,x^2\,y^3$$
$$+ K_{20}\,x^2\,y^2 + K_{21}\,x^2\,y + K_{23}\,x\,y^4 + K_{25}\,x\,y^2 + K_9\,x^4 + K_{10}\,y^4$$
$$+ K_{11}\,x^3 + K_{12}\,y^5 - K_{13}\,x^4\,y^2 - K_{14}\,x^4\,y - K_{15}\,x^3\,y^3$$
$$- K_{16}\,x^3\,y^2 - K_{17}\,x^3\,y - 2\,K_{18}\,x^2\,y^4 + 10\,K_{22}\,x\,y^5 - K_{24}\,x\,y^3$$
$$+ K_{26}\,x^2 - K_{28}\,x\,y$$

The "flattened" expression is obviously much more manageable, and perhaps is more understandable; the extraneous information contained in the constant terms has been hidden, removing some of the "clutter" in the expression. We can look at the hidden values:

```
>    K[1] = VK:-unveil( K[1] );
```

$$K_1 = 81 + 30\,z + 40\,z^7 + 88\,z^3 + z^2 - 78\,z^6 + 62\,z^5 + 11\,z^4$$

```
>    K[2] = VK:-unveil( K[2] );
```

$$K_2 = -28 + 34\,z^6 + 76\,z^3 + 65\,z^2 + 42\,z^5 - 88\,z^4 - 25\,z$$

We can verify that the new form is equivalent to the old form by subtracting the two:

```
> zero := VK:-unveil( compact1, infinity ) - p1:

> normal( zero );
```

$$0$$

We continue in the same session to demonstrate that we can have two or more sets of labels in the same session:

```
> p2 := randpoly( [r,ln(r),Y], dense, degree=6 );
```

$$
\begin{aligned}
p2 := {}& 39 - 93\,r + 6\,r\ln(r)^2\,Y^3 + 75\,Y - 28\,r^5\ln(r) + 4\,r^5\,Y \\
& - 3\ln(r)\,Y + 10\,r^4\ln(r)^2 - 82\,r^4\ln(r) - 48\,r^4\,Y^2 \\
& + 57\,r^4\ln(r)\,Y - 82\ln(r) - 5\,r^6 - 11\,r^5 + 10\ln(r)^2 - 98\,Y^2 \\
& + 38\,r^4 - 91\ln(r)^3 - 94\,Y^3 - 11\,r^4\,Y - 54\ln(r)^2\,Y \\
& + 61\ln(r)\,Y^2 - 7\,r^3\ln(r)^3 - 94\,r^3\ln(r)^2 - 35\,r^3\ln(r) \\
& - 14\,r^3\,Y^3 - 9\,r^3\,Y^2 - 51\,r^3\,Y - 83\ln(r)^3\,Y + 91\ln(r)^2\,Y^2 \\
& - 90\ln(r)\,Y^3 - 73\,r^2\ln(r)^4 + r^2\ln(r)^3 + 43\,r^2\ln(r)^2 \\
& + 67\,r^2\ln(r) - 39\,r^2\,Y^4 + 8\,r^2\,Y^3 - 49\,r^2\,Y^2 + 11\,r^2\,Y \\
& - 84\ln(r)^4\,Y - 56\ln(r)^3\,Y^2 - 93\ln(r)^2\,Y^3 - 63\ln(r)\,Y^4 \\
& - 14\,r\ln(r)^5 - 67\,r\ln(r)^4 + 76\,r\ln(r)^3 - 61\,r\ln(r)^2 \\
& - 46\,r\ln(r) - 68\,r\,Y^5 - 42\,r\,Y^4 - 47\,r\,Y^3 - 32\,r\,Y^2 + 37\,r\,Y \\
& - 90\ln(r)^5\,Y - 69\ln(r)^4\,Y^2 + 59\ln(r)^3\,Y^3 + 92\ln(r)^2\,Y^4 \\
& - 77\ln(r)\,Y^5 + 58\,r^3\ln(r)^2\,Y - 68\,r^3\ln(r)\,Y^2 + 14\,r^3\ln(r)\,Y \\
& - 91\,r^2\ln(r)^3\,Y + 5\,r^2\ln(r)^2\,Y^2 - 86\,r^2\ln(r)^2\,Y \\
& - 4\,r^2\ln(r)\,Y^3 - 50\,r^2\ln(r)\,Y^2 + 50\,r^2\ln(r)\,Y - 99\,r\ln(r)^4\,Y \\
& + 68\,r\ln(r)^3\,Y^2 + 45\,r\ln(r)^3\,Y + 72\,r\ln(r)^2\,Y^2 \\
& - 28\,r\ln(r)^2\,Y - 59\,r\ln(r)\,Y^4 + 6\,r\ln(r)\,Y^3 - 87\,r\ln(r)\,Y^2 \\
& + 72\,r\ln(r)\,Y - 73\,r^3 + 46\ln(r)^4 + 21\,Y^4 + 93\,r^2 - 53\ln(r)^5 \\
& - 40\,Y^5 - 58\ln(r)^6 + 16\,Y^6
\end{aligned}
$$

We hide these under the label C:

```
> compact2 := collect( p2, [r,ln(r)],
>     distributed, VC:-veil );
```

$$
\begin{aligned}
compact2 := {}& -C_7\,r^2\ln(r)^3 + C_{13}\,r^4\ln(r) + C_{16}\,r^2\ln(r)^2 - 28\,r^5\ln(r) \\
& + 10\,r^4\ln(r)^2 - 5\,r^6 - 7\,r^3\ln(r)^3 - 73\,r^2\ln(r)^4 - 14\,r\ln(r)^5 \\
& - 58\ln(r)^6 - C_2\,r - C_4\,r^4 - C_5\,r^3 - C_6\,r^2 - C_8\ln(r) \\
& - C_{11}\ln(r)^4 - C_{12}\ln(r)^5 + C_3\,r^5 + C_1 + C_9\ln(r)^2 + C_{10}\ln(r)^3 \\
& + 2\,C_{14}\,r^3\ln(r)^2 - C_{15}\,r^3\ln(r) - C_{17}\,r^2\ln(r) - C_{18}\,r\ln(r)^4 \\
& - C_{21}\,r\ln(r) + C_{19}\,r\ln(r)^3 + C_{20}\,r\ln(r)^2
\end{aligned}
$$

That form is much shorter, as in the last example.

```
>   zero := VC:-unveil( compact2 ) - p2:
```

```
>   expand( zero );
```

$$0$$

We can even mix the two sequences:

```
>   p3 := randpoly( [x,y,z,r,ln(r)], dense, degree=3 );
```

$$
\begin{aligned}
p3 := {} & 63 + 31\,r + 39\,x - 80\,y\,z + 8\,y^2\,z + 81\,y\,z^2 + 63\,y + 95\,z \\
& + 45\,y^2 + 85\,z^2 - 67\,y^3 - 24\,z^3 + 95\,x^3 - 68\,x^2\,y + 98\,x^2\,z \\
& + 92\,x\,y^2 - 67\,z\,r\,\ln(r) + 8\,x\,y\,z + 8\,x^2 + 44\,x\,y + 66\,x\,z^2 \\
& + 68\,x\,z - 24\,\ln(r) + 46\,\ln(r)^2 + 65\,\ln(r)^3 - 18\,r^2\,\ln(r) \\
& - 20\,r\,\ln(r)^2 + 52\,r\,\ln(r) + 95\,r^3 + 46\,r^2 - 36\,x^2\,r \\
& - 95\,x^2\,\ln(r) + 34\,y\,r - 95\,y\,\ln(r) + 19\,z\,r + 60\,z\,\ln(r) \\
& - 95\,x\,y\,r - 18\,x\,y\,\ln(r) - 62\,x\,z\,r + 40\,x\,z\,\ln(r) \\
& + 68\,x\,r\,\ln(r) + 8\,y\,z\,r - 44\,y\,z\,\ln(r) + 23\,y\,r\,\ln(r) - 67\,x\,r^2 \\
& - 65\,x\,r + 43\,x\,\ln(r)^2 + 6\,x\,\ln(r) + 20\,y^2\,r + 93\,y^2\,\ln(r) \\
& - 5\,y\,r^2 - 81\,y\,\ln(r)^2 - 63\,z^2\,r - 36\,z^2\,\ln(r) + 35\,z\,r^2
\end{aligned}
$$

```
>   compact3 := collect( p3, [x,y], distributed, VK:-veil );
```

$$
\begin{aligned}
compact3 := {} & K_{29} - K_{30}\,x - K_{31}\,y + K_{32}\,y^2 - 67\,y^3 + 95\,x^3 - 68\,x^2\,y \\
& + 92\,x\,y^2 - K_{33}\,x^2 - K_{34}\,x\,y
\end{aligned}
$$

```
>   compact4 := collect( compact3, [y], VC:-veil );
```

$$compact4 := -67\,y^3 + C_{22}\,y^2 - C_{23}\,y + C_{24}$$

```
>   VC:-unveil( C[22] );
```

$$92\,x + K_{32}$$

```
>   VK:-unveil( K[32] );
```

$$8\,z + 20\,r + 45 + 93\,\ln(r)$$

Now let us look at the computation sequence from `compact3`:

```
>   L := [seq( K[i]=VK:-unveil(K[i]), i=29..34 )];
```

$$
\begin{aligned}
L := [\,K_{29} = {} & 63 + 31\,r + 52\,r\,\ln(r) + 95\,z + 95\,r^3 + 19\,z\,r - 24\,\ln(r) \\
& - 24\,z^3 + 65\,\ln(r)^3 + 85\,z^2 - 36\,z^2\,\ln(r) - 67\,z\,r\,\ln(r) \\
& + 46\,\ln(r)^2 - 18\,r^2\,\ln(r) + 46\,r^2 + 60\,z\,\ln(r) + 35\,z\,r^2 \\
& - 63\,z^2\,r - 20\,r\,\ln(r)^2, \\
K_{30} = {} & -6\,\ln(r) - 68\,r\,\ln(r) + 62\,z\,r \\
& - 66\,z^2 - 40\,z\,\ln(r) - 68\,z + 65\,r + 67\,r^2 - 39 - 43\,\ln(r)^2,
\end{aligned}
$$

$$K_{31} = -81\,z^2 - 34\,r + 80\,z - 63 + 95\ln(r) - 8\,z\,r + 81\ln(r)^2$$
$$+ 44\,z\ln(r) + 5\,r^2 - 23\,r\ln(r),$$
$$K_{32} = 20\,r + 8\,z + 45 + 93\ln(r),$$
$$K_{33} = -8 + 36\,r - 98\,z + 95\ln(r),$$
$$K_{34} = -8\,z - 44 + 95\,r + 18\ln(r)]$$

Now, if we desire, we can convert the elements of the computation sequence to another language, say FORTRAN or C:

```
>   codegen[fortran](L,optimized);
```

```
t2 = log(r)
t3 = r*t2
t6 = r**2
t9 = z*r
t12 = z**2
t15 = t2**2
t27 = z*t2
K(29) =63+31*r+52*t3+95*z+95*t6*r+19*t9-24*t2-24*t12*z+65*|
#85*t12-36*t12*t2-67*t9*t2+46*t15-18*t6*t2+46*t6+60*t27+35*|
#t12*r-20*r*t15                                            |
K(30) =-6*t2-68*t3+62*t9-66*t12-40*t27-68*z+65*r+67*t6-39-|
t47 = 95*t2                                                |
K(31) = -81*t12-34*r+80*z-63+t47-8*t9+81*t15+44*t27+5*t6-2|
t54 = 8*z
K(32) = 20*r+t54+45+93*t2
K(33) = -8+36*r-98*z+t47
K(34) = -t54-44+95*r+18*t2
```

I remark that codegen[fortran] is careful to avoid introducing integer divides, and has decided above that it is harmless to leave the integer coefficients as integers.

```
>   codegen[C](L,optimized);
```

```
t2 = log(r);
t3 = r*t2;
t6 = r*r;
t9 = z*r;
t12 = z*z;
t15 = t2*t2;
t27 = z*t2;
K[28] =63.0+31.0*r+52.0*t3+95.0*z+95.0*t6*r+19.0*t9-24.0*t|
       65.0*t15*t2+85.0*t12-36.0*t12*t2-67.0*t9*t2+46.0*t1|
       46.0*t6+60.0*t27+35.0*z*t6-63.0*t12*r-20.0*r*t15;  |
K[29] =-6.0*t2-68.0*t3+62.0*t9-66.0*t12-40.0*t27-68.0*z+65|
       -39.0-43.0*t15;                                    |
t47 = 95.0*t2;                                            |
K[30] =-81.0*t12-34.0*r+80.0*z-63.0+t47-8.0*t9+81.0*t15+44|
       -23.0*t3;
t54 = 8.0*z;
K[31] = 20.0*r+t54+45.0+93.0*t2;
K[32] = -8.0+36.0*r-98.0*z+t47;
K[33] = -t54-44.0+95.0*r+18.0*t2;
```

I chopped the long lines output from each of those commands so that the output
would fit on this page.

2.1.3 `factor`

The Maple command `factor` is remarkably efficient at factoring multivariate
polynomials over the integers and occasionally over other fields. It is also remarkable how often in applied problems nontrivial factorings occur, and how useful it
is to find them. For example, consider the following Maple session fragment. For
simpler examples of factoring, see `?factor`.

```
>   restart;
```

```
>   with( LinearAlgebra ):
```

The following is MATLAB's gallery(3) matrix (see `help gallery` in MATLAB
for a discussion of its interesting gallery of examples).

```
>   with(Matlab):
```

```
>   evalM( "A = gallery(3)" );
```

```
>   A := map( round, getvar( "A" ) );
```

$$A := \begin{bmatrix} -149 & -50 & -154 \\ 537 & 180 & 546 \\ -27 & -9 & -25 \end{bmatrix}$$

If you do not have MATLAB on your system, you can type the matrix in directly,
as follows. The MATLAB link from Maple is useful for exploration of the special
matrices in MATLAB's gallery, though.

```
>   A := Matrix( [[-149,-50,-154],
>               [537,180,546],
>               [-27,-9,-25]] );
```

$$A := \begin{bmatrix} -149 & -50 & -154 \\ 537 & 180 & 546 \\ -27 & -9 & -25 \end{bmatrix}$$

The next matrix is about the same size as A, and chosen from the left and right
eigenvectors of A so as to produce the maximum effect on the eigenvalues of
a perturbation of A.

```
>   E := Matrix( [[130, -390, 0],
>               [43, -129, 0],
>               [133,-399,0]] );
```

$$E := \begin{bmatrix} 130 & -390 & 0 \\ 43 & -129 & 0 \\ 133 & -399 & 0 \end{bmatrix}$$

In Maple 6, to form $A + tE$ where t is a scalar we had first to convert t to the matrix diag($t, t, \dots, t$), using ScalarMatrix. In Maple 7, we can simply use simplify(A + t*E).

> AtE := simplify(A + t*E) ;

$$AtE := \begin{bmatrix} -149 + 130\,t & -50 - 390\,t & -154 \\ 537 + 43\,t & 180 - 129\,t & 546 \\ -27 + 133\,t & -9 - 399\,t & -25 \end{bmatrix}$$

The characteristic polynomial of the above matrix is

> p := CharacteristicPolynomial(AtE, x);

$$p := -6 + 11\,x - 1221271\,t + 492512\,t\,x - 6\,x^2 - t\,x^2 + x^3$$

This has multiple roots when the discriminant (see [1]) in x is zero. So, we choose t to make $d = 0$, where

> d := discrim(p, x);

$$d := 4 - 5910096\,t + 1403772863224\,t^2 \\ - 477857003880091920\,t^3 + 242563185060\,t^4$$

Let us compute a numerical value for the value of t that makes p have a multiple root. See ?fsolve, or Section 2.2.2, for a discussion of numerical solution of equations.

> alfs := [fsolve(d, t, complex)];

$$alfs := [.7837924906\,10^{-6}, .1076924816\,10^{-5} - .3085446365\,10^{-5}\,I, \\ .1076924816\,10^{-5} + .3085446365\,10^{-5}\,I, .1970031041\,10^{7}]$$

> map(abs, alfs);

$[.7837924906\,10^{-6}, .3267988117\,10^{-5}, .3267988117\,10^{-5}, .1970031041\,10^{7}]$

We see that the real root is the smallest, but that there is a pair of complex conjugate roots that is almost as small. If we wish to be guaranteed that the root is accurate, we can find an interval guaranteed to contain t by using realroot, as follows.

> realroot(d, 10^(-14));

$$\left[\left[\frac{55154493}{70368744177664}, \frac{110308987}{140737488355328} \right], \right. \\ \left. \left[\frac{27725722063861685833}{140737488355328}, \frac{1386286103193084429 17}{70368744177664} \right]\right]$$

> evalf(%);

$$[[.7837924869\,10^{-6}, .7837924940\,10^{-6}], \\ [.1970031041\,10^{7}, .1970031041\,10^{7}]]$$

So there are two real roots, one small and one large. The small root is guaranteed to be in the interval $7.837924^{95}_{86} \cdot 10^{-7}$. This means that a small perturbation of the matrix A causes its eigenvalues (which are 1, 2, and 3) to collapse to one double and one single eigenvalue. This means that the eigenvalues of A are more difficult to compute numerically than those of the average matrix. Unlike polynomial rootfinding problems, most eigenvalue problems are easy to solve numerically [58]. But we have no difficulty for this small matrix if we work symbolically, as follows. Again, we use alias to allow Maple to print its output in a compact form, representing the complicated RootOf as the label t_0.

```
>  alias( t[0]=RootOf( d, t ) );
```

$$t_0$$

```
>  As := eval( AtE, t=t[0] );
```

$$As := \begin{bmatrix} -149 + 130\,t_0 & -50 - 390\,t_0 & -154 \\ 537 + 43\,t_0 & 180 - 129\,t_0 & 546 \\ -27 + 133\,t_0 & -9 - 399\,t_0 & -25 \end{bmatrix}$$

```
>  eigs := Eigenvalues( As );
```

$$eigs :=$$

$$\left[\frac{283005565738990253}{96703623305979008} - \frac{1792424167831498089735}{8791238482361728}\,t_0\right.$$

$$+ \frac{2173324307968127776111127375}{96703623305979008}\,t_0^2$$

$$\left. - \frac{110319292597811224453495}{96703623305979008}\,t_0^3\right]$$

$$\left[\frac{297216174096883795}{193407246611958016} + \frac{1792432959069980451463}{17582476964723456}\,t_0\right.$$

$$- \frac{2173324307968127776111127375}{193407246611958016}\,t_0^2$$

$$\left. + \frac{110319292597811224453495}{193407246611958016}\,t_0^3\right]$$

$$\left[\frac{297216174096883795}{193407246611958016} + \frac{1792432959069980451463}{17582476964723456}\,t_0\right.$$

$$- \frac{2173324307968127776111127375}{193407246611958016}\,t_0^2$$

$$\left. + \frac{110319292597811224453495}{193407246611958016}\,t_0^3\right]$$

The variable t_0 is a symbolic way of representing *any* root of the discriminant.

```
>  pf := eval( p, t=t[0] );
```

$$pf := -6 + 11\,x - 1221271\,t_0 + 492512\,t_0\,x - 6\,x^2 - t_0\,x^2 + x^3$$

By construction, then, that polynomial ought to factor.

```
>   factor( pf );
```

$$\frac{1}{3617330840862075776271364437466707007583905327874048}($$

$$96703623305979008\,x - 283005565738990253$$

$$+\ 19716665846146478987085\,t_0$$

$$-\ 2173324307968127777611127375\,t_0{}^2$$

$$+\ 110319292597811224\overline{53495}\,t_0{}^3)($$

$$-193407246611958016\,x + 297216174096883795$$

$$+\ 197167625497697849660\overline{93}\,t_0$$

$$-\ 2173324307968127777611127375\,t_0{}^2$$

$$+\ 110319292597811224\overline{53495}\,t_0{}^3)^2$$

We see some application of Maple's large integers in the above factoring.

As a final exploration of this example, we plot for $t \in [0, 10^{-6}]$ the values of the eigenvalues of $A + tE$. To make things easier for the numerics, we first scale the polynomial.

```
>   p2 := eval( p, t=t/1.0e6 );
```

$$p2 := -6 + 11\,x - 1.221271000\,t + .4925120000\,t\,x - 6\,x^2$$

$$-\ .1000000000\,10^{-5}\,t\,x^2 + x^3$$

```
>   algcurves[plot_real_curve]( p2,
>               x, t, view=[1..3,0..1.0] );
```

See Figure 2.2.

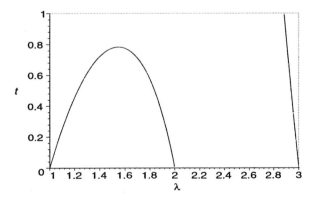

Figure 2.2: The traces of the eigenvalues of $A + 10^{-6}tE$

Discovering the eigenstructure of a matrix with parameters

The following matrix arose as a simplified example of a question that interested Kenton Yee of Brookhaven National Labs, posted to sci.math.num-analysis. We use it here to show how useful `factor` can be.

```
>  restart;
```

```
>  with( LinearAlgebra ):
```

The matrix depends on a parameter e, which cannot be zero but is otherwise unconstrained.[3]

```
>  Yee := Matrix( 8,8,
>  [[1/e, 1/e, 1/e, 1/e, 1/e, 1/e, 1/e, 0],
>  [1, 1, 1, 1, 1, 1, 0, 1],
>  [1, 1, 1, 1, 1, 0, 1, 1],
>  [1, 1, 1, 1, 0, 1, 1, 1],
>  [1, 1, 1, 0, 1, 1, 1, 1],
>  [1, 1, 0, 1, 1, 1, 1, 1],
>  [1, 0, 1, 1, 1, 1, 1, 1],
>  [0, e, e, e, e, e, e, e]]);
```

$$Yee := \begin{bmatrix} \frac{1}{e} & \frac{1}{e} & \frac{1}{e} & \frac{1}{e} & \frac{1}{e} & \frac{1}{e} & \frac{1}{e} & 0 \\ 1 & 1 & 1 & 1 & 1 & 1 & 0 & 1 \\ 1 & 1 & 1 & 1 & 1 & 0 & 1 & 1 \\ 1 & 1 & 1 & 1 & 0 & 1 & 1 & 1 \\ 1 & 1 & 1 & 0 & 1 & 1 & 1 & 1 \\ 1 & 1 & 0 & 1 & 1 & 1 & 1 & 1 \\ 1 & 0 & 1 & 1 & 1 & 1 & 1 & 1 \\ 0 & e & e & e & e & e & e & e \end{bmatrix}$$

To investigate its eigenstructure symbolically we begin by looking at its characteristic polynomial. Numerical investigations, of course, would avoid the characteristic polynomial because polynomials are much less stable than eigenvalue problems.

```
>  p := CharacteristicPolynomial( Yee, lambda );
```

$$p := (\lambda^8 e - \lambda^7 e^2 - 6\lambda^7 e - \lambda^7 + 4\lambda^6 e + 3\lambda^5 e^2 + 18\lambda^5 e + 3\lambda^5$$
$$- 18 e \lambda^4 - 3\lambda^3 e^2 - 18\lambda^3 e - 3\lambda^3 + 20\lambda^2 e + \lambda e^2 + 6\lambda e$$
$$+ \lambda - 7 e)/e$$

The polynomial itself is simplified considerably by factoring it:

---

[3] As Littlewood said, when considering the polynomial $ax^4 + bx^3 + cx^2 + dx + e$," ... here e is not necessarily the base of the natural logarithms."

```
>   factor(p);
```

$$\frac{(-1+\lambda)^3 \, (\lambda+1)^3 \, (\lambda^2 \, e - \lambda \, e^2 - 6\lambda \, e - \lambda + 7\, e)}{e}$$

The quadratic factor is of interest, because it is the only factor that depends on the parameter e.

```
>   quad_factor := normal( e*p/(lambda-1)^3/(lambda+1)^3 );
```

$$quad\_factor := \lambda^2 \, e - \lambda \, e^2 - 6\lambda \, e - \lambda + 7\, e$$

We can investigate the values of e that make the eigenstructure different, again by looking at the discriminant (see [1]).

```
>   discrim( quad_factor, lambda );
```

$$e^4 + 12\, e^3 + 10\, e^2 + 12\, e + 1$$

This does not have simple factors:

```
>   factor( % );
```

$$e^4 + 12\, e^3 + 10\, e^2 + 12\, e + 1$$

Let α be either one of the roots:

```
>   alias( alpha=RootOf(%,e) );
```

$$\alpha$$

Now with that value of e, try to factor the polynomial again:

```
>   pe := eval( p, e=alpha );
```

$$\begin{aligned} pe := (&\lambda^8\, \alpha - \lambda^7\, \alpha^2 - 6\,\lambda^7\, \alpha - \lambda^7 + 4\,\lambda^6\, \alpha + 3\,\lambda^5\, \alpha^2 + 18\,\lambda^5\, \alpha \\ &+ 3\,\lambda^5 - 18\,\alpha\,\lambda^4 - 3\,\lambda^3\, \alpha^2 - 18\,\lambda^3\, \alpha - 3\,\lambda^3 + 20\,\lambda^2\, \alpha \\ &+ \lambda\,\alpha^2 + 6\,\lambda\,\alpha + \lambda - 7\,\alpha)/\alpha \end{aligned}$$

```
>   factor( pe );
```

$$\frac{1}{4}\,(2\,\lambda + 6 + 9\,\alpha + 12\,\alpha^2 + \alpha^3)^2\,(\lambda+1)^3\,(\lambda-1)^3$$

We see that the multiplicity of the eigenvalues has changed, in that the two simple eigenvalues have coalesced. Now consider what happens if one of the roots of the quadratic factor coalesces with the root $\lambda = 1$:

```
>   resultant( quad_factor, lambda-1, lambda );
```

$$2\, e - e^2 - 1$$

```
>   factor( % );
```

$$-(e-1)^2$$

Therefore, $e = 1$ is another special value, needing separate investigation. We leave that investigation as an exercise. What about potential coalescence with $\lambda = -1$?

```
> resultant( quad_factor, lambda+1, lambda );
```
$$14\,e + e^2 + 1$$
```
> factor( % );
```
$$14\,e + e^2 + 1$$
```
> alias( beta=RootOf(%,e) );
```
$$\alpha,\ \beta$$
```
> factor( eval( p, e=beta ) );
```
$$(\lambda + 7)\,(\lambda - 1)^3\,(\lambda + 1)^4$$

So we see in that case that one triple eigenvalue becomes a quadruple eigenvalue. We do not pursue this particular problem further here, but note that the simplicity and power of factor has enabled us to identify important special cases of our problem. This success is quite typical in symbolic computation: Applied problems factor much more readily than random generic problems do, because they have structure. This structure can often be exploited.

Other abilities of factor

The routine factor can also factor over algebraic extensions of the integers. That is, if you know or suspect that the factors will contain, say, $\sqrt{2}$, you can ask Maple to factor into factors containing $\sqrt{2}$ by saying factor(poly, sqrt(2)). See ?factor for more details.

2.1.4   expand

The Maple command expand is roughly the opposite of combine, which will be discussed in Section 2.1.5. We have seen occasional examples of the use of expand in previous Maple sessions. More elementary examples can be found in the help file. The examples can be accessed directly by typing

```
> ???expand
```

and these examples show the ordinary use of expand. Using three question marks instead of one tells Maple that you wish to see only the entries under the EXAMPLES heading.

Of particular interest is the example that exhibits the so-called two-argument form of expand:

```
> expand( (x+1)*(y+z), x+1 );
```
$$(x + 1)\,y + (x + 1)\,z$$

What this has done is expand the input expression $(x + 1)(y + z)$ while keeping the subexpression $x + 1$ unexpanded. This is *opposite* to the sense of the second argument to `combine` (see Section 2.1.5), which tells `combine` to *act* on the terms containing the second argument: For `expand`, on the other hand, the second argument tells it what *not* to act on.

There is also a common use of `expand` not covered in `?expand`, when `expand` is used with `normal`. Frequently, this provides an essential part of result verification. The command `normal` usually tries to factor the numerator and denominator, and this can fail to provide a useful simplification. In that case, you pass the option `expanded` to `normal`, as follows.

```
>   normal( 1/x + 1/x^2 - 1/(x+1) + 2/(x+1), expanded );
```

$$\frac{2\,x^2 + 2\,x + 1}{x^3 + x^2}$$

We used this in the first sample session in Section 1.1.3.

2.1.5 `combine`

The Maple routine `combine` is a useful general-purpose routine for putting things together (see `?combine` for examples). I find it most useful, however, for trigonometric simplification.

```
>   restart;

>   F := cos(x)^3;
```

$$F := \cos(x)^3$$

```
>   combine( F, trig );
```

$$\frac{1}{4}\cos(3\,x) + \frac{3}{4}\cos(x)$$

This is particularly useful in perturbation calculations. One can use `sort` to replace that expression with one that puts the fundamental frequency term, $3\cos(x)/4$, first.

One can combine other functions as well, particularly logarithms. However, some of the expectations that we have are conditioned by our early exposure to the "identity" $\ln(xy) = \ln(x) + \ln(y)$, which is not true for the complex-valued logarithm. See Appendix A. Therefore, some adjustment is required in using this identity in Maple.

```
>   G := ln(x) + ln(y);
```

$$G := \ln(x) + \ln(y)$$

```
>   combine( G );
```

$$\ln(x) + \ln(y)$$

That (correctly) did nothing. Why is that correct? Because, for example, Maple has not been told that x and y are not both negative:

```
>  Why := ln(-1) + ln(-1) - ln( (-1)*(-1) );
```

$$Why := 2 I \pi$$

So, in order to get the requested transformation, we must tell Maple that the quantities are positive (or else force the transformation by giving the option symbolic to combine, an alternative that I do not recommend, because sometimes it can give incorrect results).

```
>  assume( x, positive ); assume( y, positive );
>  combine( G );
```

$$\ln(x\,y)$$
with assumptions on x and y

A more compact and manageable way to do that is with the assuming facility, new to Maple 7:

```
>  combine( ln(a) + ln(b) ) assuming positive ;
```

$$\ln(a\,b)$$

2.1.6 simplify

When all these alternatives fail or do not apply, one can try the command simplify. This routine applies various heuristics (that is, ad hoc techniques with little or no theoretical basis but that often work in practice) and can sometimes be invaluable. However, it is wise to keep a copy of the original object, because sometimes simplify will make things worse, not better.

```
>  restart;
>  e := cos(x)^5 + sin(x)^4 + 2*cos(x)^2
>           - 2*sin(x)^2 - cos(2*x);
```

$$e := \cos(x)^5 + \sin(x)^4 + 2\cos(x)^2 - 2\sin(x)^2 - \cos(2\,x)$$

```
>  simplify( e );
```

$$\cos(x)^5 + \cos(x)^4$$

The following example shows the use of simplify to prove something about continued fractions and Newton's method [42].

The continued fraction expansion for $\sqrt{2}$ is

$$1 + \cfrac{1}{2 + \cfrac{1}{2 + \cfrac{1}{2 + \ddots}}} \, ,$$

more neatly and concisely written as $1 + [2, 2, \dots]$ (this notation separates out the first entry, or partial quotient, because it alone may be zero or negative; all the other partial quotients must be positive integers). This has convergents $c_0 = 1$, $c_1 = 1 + 1/2 = 3/2, c_2 = 1 + 1/(2 + 1/2) = 7/5, c_3 = 1 + 1/(2 + 1/(2 + 1/2)) = 17/12$, $c_4 = 41/29$, $c_5 = 99/70$, $c_6 = 239/169$, $c_7 = 577/408$, and so on. Newton's method applied to the function $f(x) = x^2 - 2$ with an initial guess $x_0 = 1$ produces iterates $x_1 = 3/2, x_2 = 17/12, x_3 = 577/408$, and so on, which apparently are all convergents of the continued fraction as well.

Looking at a similar problem, we can prove that $x_k = c_{2^k - 1}$ for the positive root of $f(x) = x^2 - Nx - 1$, using Maple for the algebra in the crucial step, as follows.

```
>    restart;
```

```
>    Next := (x,N) ->  x - (x^2 - N*x - 1)/(2*x - N);
```

$$Next := (x, \, N) \rightarrow x - \frac{x^2 - N x - 1}{2 x - N}$$

That gives us the next iterate from Newton's method. It is easy to see by substituting $x = N + 1/y$ into $x^2 - Nx - 1 = 0$ (and simplifying to find that $y^2 - Ny - 1 = 0$ also) that the partial quotients (entries) in the continued fraction for the positive root x are $N + [N, N, N, \dots]$. The product of the roots is -1, so there is one positive root. If $N < 0$, we can change the equation by the substitution $x = 1/(-N + 1/y)$ to $y^2 + Ny - 1$. This gives the continued fraction $0 + [-N, -N, \dots]$ for x. We take $N > 0$ in what follows. One can then prove by induction, using the well-known recurrence relations for the convergents of a continued fraction, that each convergent in the continued fraction will be of the form $top(k)/bot(k)$, where

```
>    top := k ->  c*a^k + d*(-1/a)^k;
```

$$top := k \rightarrow c \, a^k + d \left(-\frac{1}{a} \right)^k$$

```
>    bot := k ->  e*a^k + f*(-1/a)^k;
```

$$bot := k \rightarrow e \, a^k + f \left(-\frac{1}{a} \right)^k$$

where a is defined by

```
>    N :=  a - 1/a;
```

$$N := a - \frac{1}{a}$$

We identify the constants c, d, e, and f by solving two linear systems.

```
>  solve( {top(0) = N, top(1) = N^2 + 1}, {c,d});
```

$$\left\{ c = \frac{a^3}{a^2+1},\ d = -\frac{1}{a\,(a^2+1)} \right\}$$

We can assign both c and d to be the values above with a single statement:

```
>  assign( % );
>  solve( {bot(0) = 1, bot(1) = N}, {e,f});
```

$$\left\{ e = \frac{a^2}{a^2+1},\ f = \frac{1}{a^2+1} \right\}$$

```
>  assign( % );
```

Now as an inductive step, suppose $X = \text{top}(k)/\text{bot}(k)$.

```
>  X := top(k)/bot(k);
```

$$X := \frac{\dfrac{a^3\,a^k}{a^2+1} - \dfrac{\left(-\dfrac{1}{a}\right)^k}{a\,(a^2+1)}}{\dfrac{a^2\,a^k}{a^2+1} + \dfrac{\left(-\dfrac{1}{a}\right)^k}{a^2+1}}$$

The next iterate is

```
>  nxt := Next( X, N ):
```

Now test the above for equality with c_{2k+1}. The procedure `testeq` is a *probabilistic* equality tester; it gives a high degree of confidence that two expressions are equal, and usually gives an answer much faster than a call to `simplify`.

```
>  testeq( nxt, top(2*k+1)/bot(2*k+1) );
```
                               true

This result does not provide a *proof*, but the `true` result is encouraging, and "almost" a proof. We now use `simplify` to show that they are really equal.

```
>  zero := simplify( nxt - top(2*k+1)/bot(2*k+1) );
```
                            zero := 0

This proves[4] that the result of applying a Newton iteration to any convergent $P(k)/Q(k)$ of the continued fraction for the root of $x^2 - Nx - 1$ produces the convergent $P(2k + 1)/Q(2k + 1)$. The desired result about the quadratic convergence of Newton's method follows immediately by induction.

---

[4]If you believe in `simplify`, that is.

Simplification with respect to side relations

Sometimes you will want to simplify an expression subject to some constraints, or "side relations." One can do this in Maple by a special call to `simplify`, with a second argument consisting of a set of equations to be regarded as side relations. See `?simplify[siderels]` for examples.

Finally, sometimes `simplify` gets too enthusiastic and *over*simplifies things. See the exercises.

Exercises

1. Write down three or four reasonably complicated elementary functions, such as $x^{3/5}/(1+x)$. Differentiate them using Maple; then integrate them again, and try to simplify the resulting answer to the original (plus some constant, of course).

2. Graph the functions $y_1 = \sin(\arcsin(x))$ and $y_2 = \arcsin(\sin(x))$ on $-10 \le x \le 10$. You should see graphically that y_2 is not always identically x. What does `simplify(arcsin(sin(x))` produce on your version of Maple? Why would it be wrong to produce x? Use `simplify` with the `symbolic` option.

3. Simplify $(\sin^2 x + \cos^2 x - 1)/(\tan^2 x + 1 - \sec^2 x)$ in Maple. Try again with $(\sin^2 x + \cos^2 y - 1)/(\tan^2 x + 1 - \sec^2 y)$. The first simplification is wrong in Maple 7. Is the second one?

4. Give examples to show that neither the strategy of "always expanding polynomials" nor the complementary strategy of "always factoring polynomials" is optimal for simplification. That is, find examples of polynomials that are simpler in factored form, and find others that are simpler in expanded form.

5. Expand $(\cos\theta + i\sin\theta)^5$, and simplify the output with `combine(%, trig)`. Show that `factor` can recover the input.

2.2 Solving Equations

There are a myriad of procedures in Maple to solve equations: `solve`, `fsolve`, `dsolve`, `Groebner[gsolve]`, `LinearAlgebra[LUDecomposition]`, and others. We give a brief overview here.

2.2.1 `solve`

To solve algebraic equations or systems of algebraic equations one can use the `solve` command. This is perhaps the most powerful of all Maple commands (with the possible exception of `dsolve`); though, as with many "power tools," it is possible to "cut your own leg off with it," so to speak.

```
>  restart;
```

Univariate polynomial equations of degree less than 5 can always be solved in terms of radicals. For example,

```
>  p := x^3-1;
```

$$p := x^3 - 1$$

```
>  zeros := solve( p, x );
```

$$zeros := 1, \quad -\frac{1}{2} + \frac{1}{2} I \sqrt{3}, \quad -\frac{1}{2} - \frac{1}{2} I \sqrt{3}$$

As always, we check that the results are correct.

```
>  seq( expand(eval(p,x=z)), z=zeros );
```

$$0, 0, 0$$

While solution of quartic equations in terms of radicals is also always possible, the solutions are often too complicated to be helpful; therefore, by default, Maple will return solutions of quartic equations in terms of the `RootOf` construct. To force solution in terms of radicals, you must set the environment variable `_EnvExplicit := true;` this asks Maple to use radicals even if the answer is going to be ugly. Of course, for polynomials of degree 5 or more, no general solution in terms of radicals is possible, although certain special individual polynomials may indeed have such solutions (for example, $x^5 + x + 1 = 0$). Here we look at one that is not solvable in terms of radicals.

```
>  q := x^5 - x + 1;
```

$$q := x^5 - x + 1$$

```
>  zeros := solve( q, x );
```

$$zeros := \text{RootOf}(\_Z^5 - \_Z + 1, \ index = 1),$$
$$\text{RootOf}(\_Z^5 - \_Z + 1, \ index = 2),$$
$$\text{RootOf}(\_Z^5 - \_Z + 1, \ index = 3),$$
$$\text{RootOf}(\_Z^5 - \_Z + 1, \ index = 4),$$
$$\text{RootOf}(\_Z^5 - \_Z + 1, \ index = 5)$$

```
>  seq( simplify(eval(q,x=z)), z=zeros );
```

$$0, 0, 0, 0, 0$$

Maple can also solve some systems of equations:

```
>  restart;
>  eqs := {x*t=1, x^2+t^2=4};
```
$$eqs := \{x\,t = 1,\ x^2 + t^2 = 4\}$$
```
>  soln := solve( eqs, {x,t});
```
$$soln := \{x = \mathrm{RootOf}(\_Z^4 + 1 - 4\,\_Z^2),$$
$$t = -\mathrm{RootOf}(\_Z^4 + 1 - 4\,\_Z^2)^3 + 4\,\mathrm{RootOf}(\_Z^4 + 1 - 4\,\_Z^2)\}$$
```
>  simplify( eval( eqs, soln ) );
```
$$\{4 = 4,\ 1 = 1\}$$

More powerfully, Maple can solve some transcendental equations, especially those whose solutions involve the Lambert W function [11].
```
>  solve( y*exp(y)-x, y );
```
$$\mathrm{LambertW}(x)$$
```
>  solve( y + ln(y) = x , y );
```
$$\mathrm{LambertW}(e^x)$$

The first of those equations is the definition of the Lambert W function; indeed, it defines all the branches $W_k(z)$, for all integers k. To force Maple to report all solutions, you must set the environment variable _EnvAllSolutions := true. For example, here is an equation involving arcsin:
```
>  _EnvAllSolutions := true;
```
$$\_EnvAllSolutions := true$$
```
>  solve( sin(y) - x, y );
```
$$\arcsin(x) - 2\arcsin(x)\,\_B2 + 2\pi\,\_Z2 + \pi\,\_B2$$
with assumptions on _B2 and _Z2

That solution introduced two new parameters, to parameterize the entire family of solutions of this equation. To find out about these parameters, we use the about procedure from the assume facility.
```
>  about(_B2);
```
```
Originally _B2, renamed _B2~:
  is assumed to be: OrProp(0,1)
```
```
>  about(_Z2);
```
```
Originally _Z2, renamed _Z2~:
  is assumed to be: integer
```

Imperfections of solve

The task solve sets itself is impossible. Even subject to enough restrictions to remove absolute impossibility, solution of fully general nonlinear equations is intractable. Therefore, we know at the outset that solve will fail to solve some problems. If solve thinks that it has missed some solutions, it will set the global variable SolutionsMayBeLost to be true; but detection of failure is also impossible, so sometimes even this fails. Here is an example, which may be fixed in future versions of Maple but is present in Maple 6 and Maple 7: (_EnvAllSolutions is still true, here)

```
>   solve( y=ln(exp(x)), y );
```
$$x$$

That has misfired: The unique solution is $x - 2\pi i \mathcal{K}(x)$, where the "unwinding number," discussed in Section 1.2.5, is $\mathcal{K}(z) = \lceil (\Im(z) - \pi)/(2\pi) \rceil$. See [10] for more discussion of the unwinding number, and references. Similarly,

```
>   solve( y + ln(y) = z, y );
```
$$\text{LambertW}(\_NN1, e^z)$$
$$\text{with assumptions on } \_NN1$$

We can ask about the parameter _NN1:

```
>   about( _NN1 );
```
```
Originally _NN1, renamed _NN1~:
  is assumed to be: AndProp(RealRange(0,infinity),integer)
```

This is incorrect. The solution of $y + \ln y = z$ is unique if $z \neq t \pm i\pi$ for $t \leq -1$, and is given by $y = \omega(z) := W_{\mathcal{K}(z)}(\exp z)$. On the two lines of discontinuity, $y = \omega(z)$ and $y = \omega(z - 2\pi i)$ if $z = t + i\pi$, whereas there is no such y if $z = t - i\pi$, for $t \leq -1$. See [17] for more details on this function, which we call the Wright ω function. Moreover, Maple's parameter should have been _Z1, taking on any integer value, because sometimes $\mathcal{K}(z)$ is negative. One expects future versions of Maple to get these particular equations right, but it will always be true that there will be equations that cannot be solved.

2.2.2 fsolve

To solve systems of equations numerically, one can use the fsolve (float solve) command.

```
>   restart;
>   Digits := 20;
```
$$Digits := 20$$
```
>   p := x^5-x+1;
```
$$p := x^5 - x + 1$$
```
>   r := fsolve( p, x );
```
$$r := -1.1673039782614186843$$

```
>  eval( p, x=r );
```
$$-.4\,10^{-18}$$

```
>  pseudozeros := fsolve( p, x, complex );
```

$pseudozeros := -1.1673039782614186843,$
$-.18123244446987538390 - 1.0839541013177106684\,I,$
$-.18123244446987538390 + 1.0839541013177106684\,I,$
$.76488443360058472603 - .35247154603172624932\,I,$
$.76488443360058472603 + .35247154603172624932\,I$

The command `map` maps its first argument, an operator or procedure, onto the "tuple" object (set or list or array or table) that is its second argument. It is supposed to be able to handle other arguments (after the tuple) as well, but this does not work with `evalf` because `evalf` has special evaluation rules (see `?spec_eval_rules`), and hence we use the indexed form of `evalf` to give us just one digit of accuracy (all we want is an idea of the size).

```
>  map( evalf[1],
>       [seq( abs(eval( p, x=z)), z=pseudozeros )]);
```
$$[.4\,10^{-18},\,.1\,10^{-18},\,.1\,10^{-18},\,.1\,10^{-19},\,.1\,10^{-19}]$$

This command can be used to find all complex roots of many polynomials, but complex roots of general functions are too much for it. Indeed, `fsolve` will fail on some difficult polynomials, as well (though such examples are now rare). We first set `Digits` to be the value used by hardware floats. On your system, the results may be different.

```
>  Digits := trunc( evalhf(Digits) );
```
$$Digits := 14$$

```
>  X := fsolve( x*tan(x) - 1, x );
```
$$X := -.86033358901938$$

```
>  X*tan(X)-1;
```
$$0.$$

```
>  plot( [x*tan(x),1],
>        x=-10..10, y=-5..5,
>        discont=true, colour=black );
```

That plot appears in Figure 2.3, where we see that there is a root near $x = 3$ that `fsolve` missed; but if we supply an initial guess, then `fsolve` can refine it.

```
>  X := fsolve( x*tan(x)-1, x=3);
```
$$X := 3.4256184594817$$

```
>  X*tan(X);
```
$$.99999999999990$$

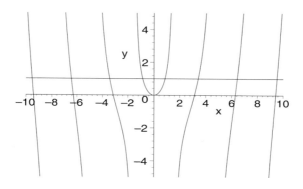

Figure 2.3: The graphs of $y = x \tan x$ and $y = 1$, superimposed

```
>   X := fsolve( x*tan(x)-1, x=6);
```
$$X := 6.4372981791719$$

```
>   X*tan(X);
```
$$.99999999999966$$

```
>   Digits := 20;
```
$$Digits := 20$$

```
>   r := fsolve( x*tan(x)-1, x=1000*Pi );
```
$$r := 3141.5929718996364202$$

```
>   r*tan(r);
```
$$.99999999999994046653$$

```
>   r - evalf(1000*Pi);
```
$$.0003183098431817$$

That number looks familiar; even more so, if we invert it.

```
>   evalf( 1/%, 7 );
```
$$3141.594$$

The following calculation has nothing to do with fsolve, but is motivated by that result coming out of fsolve (thus demonstrating one of the purposes of using fsolve in the first place, namely to try to increase our understanding). We want to do a series expansion of the large roots, near $x = n\pi$ for large integers n.

```
>  assume( n, integer );
>  T := eval( x*tan(x), x=(n*Pi+delta) );
```

$$T := (n\pi + \delta)\tan(n\pi + \delta)$$
with assumptions on n

```
>  T := series( T, delta );
```

$$T := n\pi\,\delta + \delta^2 + \frac{1}{3}\,n\pi\,\delta^3 + \frac{1}{3}\,\delta^4 + \frac{2}{15}\,n\pi\,\delta^5 + O(\delta^6)$$
with assumptions on n

Reverting that series,

```
>  solve( T=t, delta );
```

$$\frac{1}{n\pi}\,t - \frac{1}{n^3\pi^3}\,t^2 - \frac{1}{3}\frac{-6+n^2\pi^2}{n^5\pi^5}\,t^3 + \frac{1}{3}\frac{-15+4n^2\pi^2}{n^7\pi^7}\,t^4$$
$$+\frac{1}{5}\frac{70-25n^2\pi^2+n^4\pi^4}{n^9\pi^9}\,t^5 + O(t^6)$$
with assumptions on n

Now we let $t \to 1$, and we have our desired asymptotic expansion of the large roots of $x \tan x = 1$ (though we get it accurate only to $O(1/n^5)$ here, because some of the later terms affect earlier terms when $t = 1$).

`fsolve` **on systems of equations**

The routine `fsolve` uses a multidimensional version of Newton's method and so can miss roots of systems, unless given guidance in the form of initial guesses. Here is one example.

```
>  restart;
>  sys := {x^2+y^2-1,25*x*y-12};
```

$$sys := \{x^2 + y^2 - 1,\ 25\,x\,y - 12\}$$

```
>  fsolve( sys, {x,y} );
```

$$\{y = -.8000000000,\ x = -.6000000000\}$$

If `fsolve` can't find a solution, it returns unevaluated.

```
>  fsolve( sys, {x=-1..0,y=0..1} );
```

$$fsolve(\{x^2 + y^2 - 1,\ 25\,x\,y - 12\},\ \{x,\ y\},\ \{x = -1..0,\ y = 0..1\})$$

```
>  fsolve( sys, {x=0.2..1.2, y=0.2..1.2} );
```

$$\{x = .8000000000,\ y = .6000000000\}$$

```
>  fsolve( sys, {x=-1.2..-0.2, y=-1.2..-0.2} );
```

$$\{y = -.8000000000,\ x = -.6000000000\}$$

2.2.3 dsolve

One of the biggest changes in Maple since the first edition of this book is the dramatic improvement of the dsolve routine for the solution of differential equations. Previously quite powerful, it is now the most powerful differential equation solver implemented in any computer algebra system. It can solve many kinds of ordinary differential equations analytically. I give only a few examples here.

A differential equation related to Fibonacci numbers

```
>    restart;
```

```
>    fib := (-4-6*x-6*x^2)*Y(x)+(-1+x+x^2)^2*diff(Y(x),x,x);
```

$$fib := (-4 - 6x - 6x^2)\, Y(x) + (-1 + x + x^2)^2 \, (\tfrac{\partial^2}{\partial x^2} \, Y(x))$$

We will see that the power series expansion of the solution of that equation that satisfies $Y(0) = Y'(0) = 1$ has coefficients equal to the Fibonacci numbers.

```
>    Order := 14;
```

$$Order := 14$$

```
>    dsolve( {fib,Y(0)=1,D(Y)(0)=1}, Y(x), series );
```

$$Y(x) = 1 + x + 2x^2 + 3x^3 + 5x^4 + 8x^5 + 13x^6 + 21x^7 + 34x^8 + 55 x^9 + 89x^{10} + 144x^{11} + 233x^{12} + 377x^{13} + O(x^{14})$$

```
>    seq(combinat[fibonacci](n),n=1..14);
```

$$1,\ 1,\ 2,\ 3,\ 5,\ 8,\ 13,\ 21,\ 34,\ 55,\ 89,\ 144,\ 233,\ 377$$

```
>    dsolve( fib, Y(x) );
```

$$Y(x) = \frac{\_C1}{-1 + x + x^2} + \frac{\_C2\,(6x^5 + 15x^4 - 10x^3 - 30x^2 + 30x)}{-6 + 6x + 6x^2}$$

A model combustion problem

The following problem is used in [49] as an expository example to show how to use singular perturbation methods. The perturbation methods expose a boundary layer of width ε at time $O(1/\varepsilon)$, which marks the time of combustion of a substance. Here, we give the exact solution, as computed by Maple 7.

```
>    de := diff( y(x), x ) = y(x)^2*(1-y(x));
```

$$de := \tfrac{\partial}{\partial x}\, y(x) = y(x)^2\,(1 - y(x))$$

We solve it with an initial condition $\varepsilon = 10^{-k}$, where k is yet symbolic.

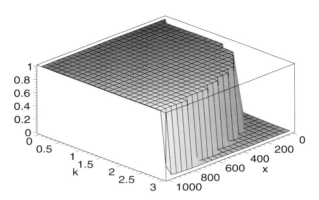

Figure 2.4: Three-dimensional plot showing the dependence of the combustion region on $\varepsilon = 10^{-k}$

```
>   sn := dsolve( {de,y(0)=10^(-k)}, y(x) );
```

$$sn := y(x) = \cfrac{1}{\text{LambertW}\left(-\cfrac{(-1 + 10^{(-k)}) \exp\left(-\frac{-1+10^{(-k)}}{10^{(-k)}}\right) e^{(-x-1)}}{e^{(-1)} 10^{(-k)}}\right) + 1}$$

The rather ugly argument to W can be simplified to

$$(10^k - 1)e^{10^k - 1 - x} ,$$

but we don't worry about making Maple do that here. We can plot the solution for a number of k-values simultaneously, showing the dependence on k of the timing of the rapid rise in x from near 0 to near 1:

```
>   plot3d( rhs(sn), x=0..10^3, k=0..3, axes=BOXED );
```

We can also examine one particular value of k, and look closely at the combustion region:

```
>   trial := eval( sn, k=5 );
```

$$trial := y(x) = \cfrac{1}{\text{LambertW}\left(99999 \cfrac{e^{99999} e^{(-x-1)}}{e^{(-1)}}\right) + 1}$$

```
>   plot( rhs(trial), x=0.999e5..1.001e5 );
```

That plot is in Figure 2.5.

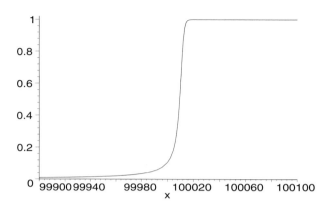

Figure 2.5: A blowup of the combustion region for $\varepsilon = 10^{-5}$

The Lotka–Volterra equations

The following is a special case of the Lotka–Volterra equations [4]. This is also Problem B1 of the DETEST suite [33].

```
>   dsys := {diff(x(t),t)=x(t)*(1-y(t)),
>               diff(y(t),t)=s*y(t)*(x(t)-1) };
```

$$dsys := \{\tfrac{\partial}{\partial t} x(t) = x(t)\,(1 - y(t)),\ \tfrac{\partial}{\partial t} y(t) = s\,y(t)\,(x(t) - 1)\}$$

```
>   de := diff( x(y), y ) =
>   eval( rhs(dsys[1])/rhs(dsys[2]), [x(t)=x(y), y(t)=y] );
```

$$de := \tfrac{\partial}{\partial y} x(y) = \frac{x(y)\,(1 - y)}{s\,y\,(x(y) - 1)}$$

```
>   _EnvAllSolutions := true;
```

$$\_EnvAllSolutions := true$$

```
>   dsolve( de, x(y) );
```

$$x(y) = \exp\left(\frac{y - s\,\mathrm{LambertW}(\_N N1,\ -y^{(-\frac{1}{s})}\,\exp(\frac{y}{s} + \frac{\_C1}{s} - \frac{2\,I\,\pi\,\_Z2}{s})) + \_C1 - \ln(y) - 2\,I\,\pi\,\_Z2}{s}\right)$$

with assumptions on $\_NN1$, $\_Z2$

I modified that output (changing e^a to $\exp a$) to make it slightly more legible.

```
>   expand( % ):
>   simplify( % );
```

$$x(y) = -\mathrm{LambertW}(\_N N1,\ -y^{(-\frac{1}{s})}\,\exp(\tfrac{y + \_C1 - 2\,I\,\pi\,\_Z2}{s}))$$

with assumptions on $\_NN1$, $\_Z2$

Inspection of that solution shows that it is real-valued for nonzero $\_Z2$ only if s is ± 1. Later we take $s \approx 0.3$ and so we may remove the $\_Z2$ as follows.

```
>  eval( %, _Z2=0 );
```

$$x(y) = -\text{LambertW}(\_NN1, -y^{(-\frac{1}{s})} \exp(\tfrac{y+\_C1}{s}))$$
with assumptions on $\_NN1$

The only branches of W_k that have real values are W_0 and W_{-1}. So we specify this as follows.

```
>  x1 := eval( %, _NN1=0 );
```

$$x1 := x(y) = -\text{LambertW}(0, -y^{(-\frac{1}{s})} e^{(\frac{y+\_C1}{s})})$$

```
>  x2 := eval( %%, _NN1=-1 );
```

$$x2 := x(y) = -\text{LambertW}(-1, -y^{(-\frac{1}{s})} e^{(\frac{y+\_C1}{s})})$$

```
>  s := rand()/1.0e12;
```

$$s := .2726006090$$

```
>  toplot := { [rhs(x1),y,y=0..10], [rhs(x2),y,y=0..10] };
```

$$toplot := \{[-\text{LambertW}(-\frac{e^{(3.668370381\,y+3.668370381\,\_C1)}}{y^{3.668370381}}), \, y, \, y = 0..10],$$

$$[-\text{LambertW}(-1, \, -\frac{e^{(3.668370381\,y+3.668370381\,\_C1)}}{y^{3.668370381}}), \, y, \, y = 0..10]\}$$

```
>  sols := 'union'( seq(toplot,
>           _C1={-7,-6,-5,-4,-3,-2,-3/2,-1-s}) ):
>  plot( %, view=[0..20,0..10], colour=black );
```

See Figure 2.6. The branches do not quite match up because it is difficult to plot adaptively near a derivative singularity (which these are because we are plotting x as a function of y); if we superimposed the equivalent solution for $y(x)$ that could be obtained by solving the inverse equation, we would see that the phase portrait consists of closed curves.

```
>  eval( [x1,x2], y=1 );
```

$$[x(1) = -\text{LambertW}(-1.\,e^{(3.668370381+3.668370381\,\_C1)}),$$
$$x(1) = -\text{LambertW}(-1, \, -1.\,e^{(3.668370381+3.668370381\,\_C1)})]$$

Inspection shows that the centre at $[1, 1]$ occurs if $\_C1 = -1 - s$, as follows:

```
>  subs( _C1=-1-s, % );
```

$$\big[x(1) = -\text{LambertW}(-1.\,e^{(-1.000000000)}),$$
$$x(1) = -\text{LambertW}(-1, \, -1.\,e^{(-1.000000000)})\big]$$

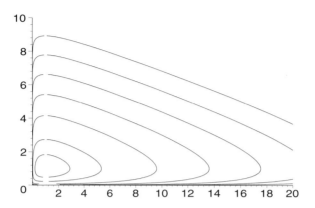

Figure 2.6: A phase portrait of the Lotka–Volterra equations, plotted from the analytic solution

>   evalf(%);

$$[x(1) = .9999999999 - .00001246016198\,I,$$
$$x(1) = .9999999999 + .00001246016198\,I]$$

As we would expect, the differences from 1, which are due to rounding errors and are about 10^{-5}, are about the square root of the errors in the arguments (which are about 10^{-10}) because this is a second-order singularity.

>   s := 's';

$$s := s$$

>   solv   X1=rhs(eval(x1,y=1)), _C1);

$$-1 + s\ln(X1) - s\,X1 + 2\,I\,s\,\pi\,\_Z5\tilde{}$$

>   eval(%, _Z5=0);

$$-1 + s\ln(X1) - s\,X1$$

>   eval(dsys, s=0);

$$\{\tfrac{\partial}{\partial t}\,y(t) = 0,\ \tfrac{\partial}{\partial t}\,x(t) = x(t)\,(1 - y(t))\}$$

>   dsolve(%, {x(t),y(t)});

$$[\{y(t) = \_C2\},\ \{x(t) = \_C1\,e^{(\int 1 - y(t)\,dt)}\}]$$

```
>   dsolve(dsys,{x(t),y(t)});
```

$$\left[\{x(t) = 0\}, \{y(t) = \_C1\, e^{(-s\,t)}\}\right], \left[\left\{x(t) = \text{RootOf}\left(\vphantom{\int}\right.\right.\right.$$

$$\int^{\_Z}_{\_a} \frac{1}{\_a\left(\text{LambertW}\left(\_NN2, \dfrac{e^{(s\,\_a)}\, e^{(\_C1\,s)}\, e^{(-1)}}{\_a^s\, (e^{(I\,s\,\pi\,\_Z7)})^2}\right) + 1\right)}\, d\_a - t$$

$$\left.\left.\left. - \_C2\right)\right\}, \left\{y(t) = \frac{-(\frac{\partial}{\partial t}\, x(t)) + x(t)}{x(t)}\right\}\right]$$

with assumptions on $\_NN2$ and $\_Z7$

That complicated representation of the exact solution can be manipulated in Maple, albeit not as easily as a completely explicit solution. It essentially shows that the differential equation can be reduced to quadrature.

The Airy equation, and an example of Hubbard and West

We begin with a reminder of the Airy functions $Ai(x)$ and $Bi(x)$. They are defined by the differential equation

```
>   restart;
>   AiryDE := diff( y(z), z, z ) = z*y(z);
```

$$AiryDE := \frac{\partial^2}{\partial z^2}\, y(z) = z\, y(z)$$

```
>   dsolve( AiryDE, y(z) ) ;
```

$$y(z) = \_C1\, \text{AiryAi}(z) + \_C2\, \text{AiryBi}(z)$$

and certain initial conditions are given, so that $Ai(x) \to 0$ as $x \to \infty$, whereas the asymptotics of $Bi(x)$ and its derivative $Bi'(x)$ (denoted by `AiryBi(1,x)` in Maple) are

```
>   asympt( AiryBi(1,t), t, 2 );
```

$$\frac{e^{(2/3\,t^{(3/2)})}}{\sqrt{\pi}\left(\frac{1}{t}\right)^{(1/4)}} - \frac{7}{48}\frac{e^{(2/3\,t^{(3/2)})}\left(\frac{1}{t}\right)^{(5/4)}}{\sqrt{\pi}} + O\left(\left(\frac{1}{t}\right)^{(11/4)}\right)$$

```
>  asympt( AiryBi(t), t, 2 );
```

$$\frac{e^{\left(2/3\,t^{(3/2)}\right)}\left(\dfrac{1}{t}\right)^{(1/4)}}{\sqrt{\pi}} + \frac{\dfrac{5}{48}\,e^{\left(2/3\,t^{(3/2)}\right)}\left(\dfrac{1}{t}\right)^{(7/4)}}{\sqrt{\pi}} + O\left(\left(\frac{1}{t}\right)^{(13/4)}\right)$$

Moreover, the asymptotics of the ratio $\mathrm{Bi}'(x)/\mathrm{Bi}(x)$ are given by

```
>  asympt( %% / %, t, 2 );
```

$$\frac{1}{\sqrt{\dfrac{1}{t}}} - \frac{1}{4}\frac{1}{t} + O\left(\left(\frac{1}{t}\right)^{(5/2)}\right)$$

to high enough order to be useful later. Note that Maple has not simplified $\sqrt{1/t}$ to $1/\sqrt{t}$, because we have not told it explicitly that $t > 0$. See Appendix A. We now consider an example from [32], one of my favourite differential equations texts:

```
>  HubbardWest := diff( x(t), t ) = x(t)^2 - t;
```

$$\textit{HubbardWest} := \frac{\partial}{\partial t}\,\mathrm{x}(t) = x(t)^2 - t$$

Maple can solve this differential equation analytically:

```
>  dsolve( HubbardWest, x(t) );
```

$$x(t) = -\frac{\_C1\,\mathrm{AiryAi}(1,\,t) + \mathrm{AiryBi}(1,\,t)}{\_C1\,\mathrm{AiryAi}(t) + \mathrm{AiryBi}(t)}$$

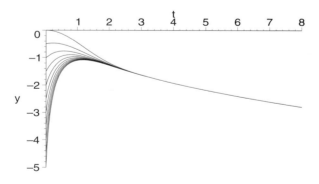

Figure 2.7: The exact solution of $\dot{x} = x^2 - t$ for various initial conditions

```
> dsolve( {HubbardWest, x(0)=alpha}, x(t) );
```

$$x(t) = -\cfrac{-\cfrac{(3\,\Gamma(\tfrac{2}{3})^2\,3^{(2/3)} + 2\,\alpha\,\pi\,3^{(5/6)})\,\mathrm{AiryAi}(1,\,t)}{-3\,\Gamma(\tfrac{2}{3})^2\,3^{(1/6)} + 2\,\alpha\,\pi\,3^{(1/3)}} + \mathrm{AiryBi}(1,\,t)}{-\cfrac{(3\,\Gamma(\tfrac{2}{3})^2\,3^{(2/3)} + 2\,\alpha\,\pi\,3^{(5/6)})\,\mathrm{AiryAi}(t)}{-3\,\Gamma(\tfrac{2}{3})^2\,3^{(1/6)} + 2\,\alpha\,\pi\,3^{(1/3)}} + \mathrm{AiryBi}(t)}$$

We will now plot that solution for several different initial conditions.

```
> X := rhs( % ):
> Nplot := 10;
```

$$Nplot := 10$$

```
> P := plot( [seq(X,alpha=[seq(-5*i/Nplot, i=0..Nplot)])],
>         t=0..8, y=-5..0,
>         scaling=CONSTRAINED ): plots[display](P);
```

We now look at a bad numerical solution, "designed to fool people who trust computers" [32]. The problem with this numerical solution is that it is from a fixed step size method, with no error estimation or adaptive step-size control. We choose to use a classical fourth order Runge–Kutta method ("the" Runge–Kutta method, as it is often erroneously called) with fixed step size $h = 0.1$. The step size is, of course, too large to get accurate results; but it is a venerable step size often used in ignorance. Here is one example of what can go wrong.

```
> P1 := plot( eval(X,alpha=0),
>         linestyle=4, t=0..300 ): plots[display](P1);
```

That plot is not shown by itself, but is included in Figure 2.8.

```
> N := 3000;
```

$$N := 3000$$

```
> h := 300.0/N;
```

$$h := .1000000000$$

```
> badsol :=
> dsolve({HubbardWest,x(0)=0},
>         numeric,
>         method=classical[rk4],
>         stepsize=h,
>         output=procedurelist);

> wrongpts := [seq( subs( badsol(k*h),[t,x(t)]), k=0..N)]:
> P2 := plot( wrongpts, style=POINT,
>         colour=black, symbol=POINT ):
> plots[display]( {P,P2} );
```

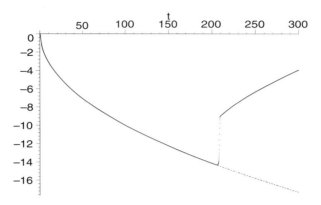

Figure 2.8: Comparison of a bad numerical solution to the exact solution

Plotting those on the same graph (see Figure 2.8), we see dramatically spurious behaviour from the fixed time-step $h = 0.1$ solution by classical fourth-order Runge-Kutta. This kind of bad behaviour is why you should use better methods, such as exemplified below.

Now, for comparison, we solve this problem with a good numerical method (which is not only accurate, but also fast), the stiff solver from [52], written by Larry Shampine and incorporated into Maple 7 by Allan Wittkopf. [The problem gets stiffer as x gets larger.]

> A problem is "stiff" if in comparison
> ode15s mops the floor with ode45.
> —Larry Shampine

> Stiffness is a confluence of a problem,
> a method, and a user's expectations.
> —Arieh Iserles

Arieh's remark summarizes a great deal of technical and theoretical work; it is concise, correct, comprehensive, and helpful in explaining why stiff methods work or don't work. As Larry says, though, the practical (though circular) definition of a stiff problem is one for which a stiff method performs better; one tries a nonstiff method first, and if it appears to bog down, one tries again with a stiff method. One doesn't try the stiff method first (without good reason) because non-stiff problems are more common, and stiff methods can be relatively inefficient for nonstiff problems. See ?stiffness.

```
>   goodsol :=
>   dsolve( {HubbardWest,x(0)=0},
>        numeric,
>        stiff=true,
>        range=0..10^8 );
```

$$goodsol := \mathbf{proc}(rosenbrock\_x) \ldots \mathbf{end\ proc}$$

```
>   plots[odeplot]( goodsol, style=POINT,
>            symbol=POINT, colour=BLACK );
```

See Figure 2.9. The solution method is a low-order Rosenbrock method with adaptive stepsize control, suitable for stiff systems. This solution can be compared with the exact solution:

```
>   goodsol( 1.0e5 );
```

$$[t = 100000., \; x(t) = -316.227807972679954]$$

```
>   eval( X, [alpha=0,t=1.0e5] );
```

$$-\frac{-.1298219251\,10^{-9155730}\,\sqrt{3} + .3876788305\,10^{9155733}}{.4105329705\,10^{-9155733}\,\sqrt{3} + .1225948115\,10^{9155731}}$$

Notice the extremely large numbers that appear in that ratio: we will see that this causes problems for the exact solution.

```
>   evalf( % );
```

$$-316.2277634$$

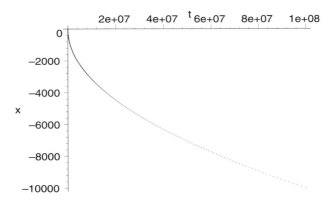

Figure 2.9: A good numerical solution of $\dot{x} = x^2 - t$, $x(0) = 0$

Now try a larger value of x in the numerical solution:

```
>   goodsol( 1.0e7 );
```

$$[t = .10\,10^8,\ x(t) = -3162.27779889634122]$$

Maple did tell us that the asymptotics were $x \sim -\sqrt{t}$, and we see the following:

```
>   sqrt(1.0e7);
```

$$3162.277660$$

Now try the analytical solution at that value of x:

```
>   eval( X, [alpha=0,t=1.0e7] );
```

$$\text{Float(undefined)}$$

Maple was unable to evaluate the analytical solution in terms of Airy functions to this problem, for large x. But the numerical solution works for even larger x:

```
>   goodsol( 1.0e8 );
```

$$[t = .10\,10^9,\ x(t) = -9999.9999032634078]$$

The moral that can be derived from this is that even if you have an analytical answer, you have to use it carefully.

Another problem needing numerical solution.

The following differential equation was posed by Mehmet Ali Suzen to the Maple User's Group. We explore its numerical solution, for a large number of initial conditions, as follows. We build a procedure that evaluates the differential equation for N different initial conditions at once (here $m = 4N$ because the original differential equation was two second order differential equations, so each group of 4 corresponds to one solution of the original equation with its own initial condition).

   Remark. In contrast to the practice in MATLAB, you must *not* specify storage for the vector of derivatives: Maple will handle it for you, and indeed if you try, you will slow the computation down.

```
>   restart;
>   Suzen := proc (m, t, yvec, ypvec)
>       local i, j;
>       option '[yvec[1] = x(t), yvec[2] = diff(x(t),t),
>          yvec[3] = y(t), yvec[4] = diff(y(t),t)]';
>       for i to m/4 do
>          j := 4*(i-1);
>          ypvec[1+j] :=   yvec[2+j];
>          ypvec[2+j] :=  -yvec[1+j]^2+yvec[3+j]^2;
>          ypvec[3+j] :=   yvec[4+j];
>          ypvec[4+j] := 2*yvec[1+j]*yvec[3+j];
>       end do;
>       ypvec
>   end proc:
```

```
>   N := 72;
```

$$N := 72$$

The initial conditions are points equally spaced around the unit circle, with zero velocity.

```
>   ic := map( evalf, vector( 4*N,
>   [seq(op(
>           [ cos(2*Pi*(i-1)/N), 0,
>               sin(2*Pi*(i-1)/N),0]),
>           i=1..N ) ] ) ):
```

The numerical solution proceeds using the method described in [52], as incorporated into Maple 7 by Allan Wittkopf.

```
>   st := time():
>   sol := dsolve( numeric, procedure=Suzen,
>           range=0..3,
>           start=0, initial=ic,
>           procvars=[seq(z||i(t),i=1..4*N)] ):
>   time_taken_de := time() - st;
```

```
Warning, cannot evaluate the solution further right of 2.9744778,
probably a singularity
```

$$time\_taken\_de := 2.303$$

Of course the time taken will be different on your machine. That singularity seems genuine, and may represent a path crossing (in 4-space—the picture we see below is only a two-dimensional section).

Unfortunately, the plotting portion of the code reported in [52] has not yet been incorporated into Maple, and so plotting the solution (computed so quickly) takes unnecessarily more time than the solution does. Interim code for fast plotting of such solutions may be available before the next version of Maple.

Note the use of labels=["",""] in the following to turn off the printing of the legend. See Figure 2.10 for the plot. We have specified the variables to use in this command by using the procvars option in the call to dsolve. What we are doing is plotting the first and third variables of each group of four against each other—that is, we are plotting a phase portrait for the original differential equation, with a solution arising from each of the 72 different initial conditions.

```
>   st := time():
>   plots[odeplot]( sol,
>       [seq([z||(1+4*(i-1))(t),z||(3+4*(i-1))(t)],i=1..N)],
>       colour=black, view=[-3..3,-3..3],
>       scaling=CONSTRAINED, labels=["",""], axes=BOXED );
>   time_taken_plotting := time() - st;
```

$$time\_taken\_plotting := 50.984$$

See Figure 2.10.

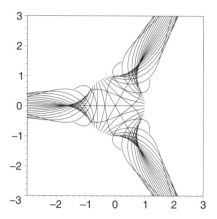

Figure 2.10: Graph of 72 solutions of a pair of second-order equations

2.2.4 rsolve

One often wants to solve *finite-difference equations* (also known as *recurrence relations*). This is done in Maple with the `rsolve` command. This can solve some linear constant-coefficient recurrence equations, some linear polynomial-coefficient recurrence equations, and some divide-and-conquer-type recurrence equations that arise in the analysis of algorithms. For example, suppose that your implementation of Strassen's algorithm [55] for fast matrix multiplication of n-by-n matrices divides the computation up into 7 multiplications of $(n/2)$-by-$(n/2)$ matrices, at the cost of 15 additions of $(n/2)$-by-$(n/2)$ matrices plus $3n^2$ addressing operations, which you assume have the same cost as a floating-point addition, and that the cost of the 1-by-1 base case is a floating-point multiply, which you assume costs as much as 5 floating-point additions. The recurrence equation you get, then, can be expressed in Maple as follows.

```
>   restart;
>   req := {Strassen(n) = 7*Strassen(n/2)
>          + 15*(n/2)^2 + 3*n^2,
>       Strassen(1)=5};
```

$$req := \{Strassen(n) = 7\,Strassen(\frac{1}{2}\,n) + \frac{27}{4}\,n^2,\ Strassen(1) = 5\}$$

We ask Maple to solve this equation by using `rsolve`:

```
>   rsolve( req, Strassen(n) );
```

$$5\,n^{\left(\frac{\ln(7)}{\ln(2)}\right)} + n^{\left(\frac{\ln(7)}{\ln(2)}\right)}\left(-\frac{63}{4}\,(\frac{4}{7})^{\left(\frac{\ln(n)}{\ln(2)}+1\right)} + 9\right)$$

```
>  factor( % );
```

$$-\frac{7}{4}\, n^{\left(\frac{\ln(7)}{\ln(2)}\right)} \left(-8 + 9\,(\frac{4}{7})^{\left(\frac{\ln(n)+\ln(2)}{\ln(2)}\right)}\right)$$

```
>  (-7/4)*(-8);
```

$$14$$

That last computation was just to draw your attention to where the factor 14 comes from in the analysis of this result that follows.

A little thought shows that the $(4/7)^{\log n}$ factor goes to zero as $n \to \infty$, so the cost is asymptotically $14n^{2.807\cdots}$, since $\ln(7)/\ln(2) = 2.807\ldots$, less than the $O(n^3)$ cost of standard matrix multiplication. The optimal implementation stops at 8-by-8 blocks, and has a cost $4n^{\ln 7/\ln 2}$ [31, p. 447], and so this implementation isn't so good.

2.2.5 Linear Equations

Systems of linear algebraic equations can be solved many ways in Maple. For sparse systems of linear equations expressed in matrix form, the routine LinearSolve is efficient and simple. For some problems with symbolic parameters, though, the Turing factoring, as implemented in LUDecomposition, is more reliable (other routines such as ReducedRowEchelonForm may miss solutions) [15]. We give an example here of the use of Turing factoring, namely on the matrix from Kenton Yee mentioned earlier. Note that this matrix is not defined for $e = 0$.

```
>  restart;
>  with( LinearAlgebra ):
>  Yee := Matrix( 8,8,
>  [[1/e, 1/e, 1/e, 1/e, 1/e, 1/e, 1/e, 0],
>  [1, 1, 1, 1, 1, 1, 0, 1],
>  [1, 1, 1, 1, 1, 0, 1, 1],
>  [1, 1, 1, 1, 0, 1, 1, 1],
>  [1, 1, 1, 0, 1, 1, 1, 1],
>  [1, 1, 0, 1, 1, 1, 1, 1],
>  [1, 0, 1, 1, 1, 1, 1, 1],
>  [0, e, e, e, e, e, e, e]]);
```

$$Yee := \begin{bmatrix} \frac{1}{e} & \frac{1}{e} & \frac{1}{e} & \frac{1}{e} & \frac{1}{e} & \frac{1}{e} & \frac{1}{e} & 0 \\ 1 & 1 & 1 & 1 & 1 & 1 & 0 & 1 \\ 1 & 1 & 1 & 1 & 1 & 0 & 1 & 1 \\ 1 & 1 & 1 & 1 & 0 & 1 & 1 & 1 \\ 1 & 1 & 1 & 0 & 1 & 1 & 1 & 1 \\ 1 & 1 & 0 & 1 & 1 & 1 & 1 & 1 \\ 1 & 0 & 1 & 1 & 1 & 1 & 1 & 1 \\ 0 & e & e & e & e & e & e & e \end{bmatrix}$$

We find the Turing factoring as follows.

```
>    ( P,L,U,R,det ) :=
>            LUDecomposition( Yee,
>                output=['P','L','U1','R','determinant'] ):
```

Note the assignment to each element of an expression sequence of names. This works only if the the number of elements matches the size of the expression sequence on the right. Here, P is a permutation matrix, L is unit lower-triangular, U is upper-triangular, R is the unique reduced row-echelon form of the input matrix (in this case R is just the identity, because the Yee matrix is nonsingular for all nonzero values of the parameter e). This is not quite the Turing factoring $PA = LDUR$ where the permutation matrix goes on the other side and the extra matrix D is diagonal, but it is close enough. We use the Norm function to compute a simple matrix norm, so we don't have to print a matrix of zeros.

```
>    Norm( Yee-P.L.U.R, infinity );
```
$$0$$

The most important thing is that the matrix R is *guaranteed to be the unique reduced row-echelon form of A* only if the determinant of $U_1 = DU$ is nonzero. This is an example of what is called a *proviso* on an answer returned by a computer algebra system.

```
>    det;
```
$$-7$$

Now let us look at this factoring.

```
>    P;
```

$$\begin{bmatrix} 1 & 0 & 0 & 0 & 0 & 0 & 0 & 0 \\ 0 & 0 & 0 & 0 & 0 & 0 & 1 & 0 \\ 0 & 0 & 0 & 0 & 0 & 1 & 0 & 0 \\ 0 & 0 & 0 & 0 & 1 & 0 & 0 & 0 \\ 0 & 0 & 0 & 1 & 0 & 0 & 0 & 0 \\ 0 & 0 & 1 & 0 & 0 & 0 & 0 & 0 \\ 0 & 1 & 0 & 0 & 0 & 0 & 0 & 0 \\ 0 & 0 & 0 & 0 & 0 & 0 & 0 & 1 \end{bmatrix}$$

```
>    L;
```

$$\begin{bmatrix} 1 & 0 & 0 & 0 & 0 & 0 & 0 & 0 \\ e & 1 & 0 & 0 & 0 & 0 & 0 & 0 \\ e & 0 & 1 & 0 & 0 & 0 & 0 & 0 \\ e & 0 & 0 & 1 & 0 & 0 & 0 & 0 \\ e & 0 & 0 & 0 & 1 & 0 & 0 & 0 \\ e & 0 & 0 & 0 & 0 & 1 & 0 & 0 \\ e & 0 & 0 & 0 & 0 & 0 & 1 & 0 \\ 0 & -e & -e & -e & -e & -e & -e & 1 \end{bmatrix}$$

```
>  U;
```

$$\begin{bmatrix} \dfrac{1}{e} & \dfrac{1}{e} & \dfrac{1}{e} & \dfrac{1}{e} & \dfrac{1}{e} & \dfrac{1}{e} & \dfrac{1}{e} & 0 \\ 0 & -1 & 0 & 0 & 0 & 0 & 0 & 1 \\ 0 & 0 & -1 & 0 & 0 & 0 & 0 & 1 \\ 0 & 0 & 0 & -1 & 0 & 0 & 0 & 1 \\ 0 & 0 & 0 & 0 & -1 & 0 & 0 & 1 \\ 0 & 0 & 0 & 0 & 0 & -1 & 0 & 1 \\ 0 & 0 & 0 & 0 & 0 & 0 & -1 & 1 \\ 0 & 0 & 0 & 0 & 0 & 0 & 0 & 7e \end{bmatrix}$$

Since those factors are all so simple, it would be too good to be true that their inverses would be, also: But it is true!

```
>  L^(-1);
```

$$\begin{bmatrix} 1 & 0 & 0 & 0 & 0 & 0 & 0 & 0 \\ -e & 1 & 0 & 0 & 0 & 0 & 0 & 0 \\ -e & 0 & 1 & 0 & 0 & 0 & 0 & 0 \\ -e & 0 & 0 & 1 & 0 & 0 & 0 & 0 \\ -e & 0 & 0 & 0 & 1 & 0 & 0 & 0 \\ -e & 0 & 0 & 0 & 0 & 1 & 0 & 0 \\ -e & 0 & 0 & 0 & 0 & 0 & 1 & 0 \\ -6e^2 & e & e & e & e & e & e & 1 \end{bmatrix}$$

```
>  U^(-1);
```

$$\begin{bmatrix} e & 1 & 1 & 1 & 1 & 1 & 1 & -\dfrac{6}{7}\dfrac{1}{e} \\ 0 & -1 & 0 & 0 & 0 & 0 & 0 & \dfrac{1}{7}\dfrac{1}{e} \\ 0 & 0 & -1 & 0 & 0 & 0 & 0 & \dfrac{1}{7}\dfrac{1}{e} \\ 0 & 0 & 0 & -1 & 0 & 0 & 0 & \dfrac{1}{7}\dfrac{1}{e} \\ 0 & 0 & 0 & 0 & -1 & 0 & 0 & \dfrac{1}{7}\dfrac{1}{e} \\ 0 & 0 & 0 & 0 & 0 & -1 & 0 & \dfrac{1}{7}\dfrac{1}{e} \\ 0 & 0 & 0 & 0 & 0 & 0 & -1 & \dfrac{1}{7}\dfrac{1}{e} \\ 0 & 0 & 0 & 0 & 0 & 0 & 0 & \dfrac{1}{7}\dfrac{1}{e} \end{bmatrix}$$

```
>  P^(-1);
```

$$
\begin{bmatrix}
1 & 0 & 0 & 0 & 0 & 0 & 0 & 0 \\
0 & 0 & 0 & 0 & 0 & 0 & 1 & 0 \\
0 & 0 & 0 & 0 & 0 & 1 & 0 & 0 \\
0 & 0 & 0 & 0 & 1 & 0 & 0 & 0 \\
0 & 0 & 0 & 1 & 0 & 0 & 0 & 0 \\
0 & 0 & 1 & 0 & 0 & 0 & 0 & 0 \\
0 & 1 & 0 & 0 & 0 & 0 & 0 & 0 \\
0 & 0 & 0 & 0 & 0 & 0 & 0 & 1
\end{bmatrix}
$$

This means that we may represent the inverse of the Yee matrix as a product of sparse matrices, allowing for $O(n)$ multiplications in solving the system $Yx = b$. This is not important for this example, where n is only 8, but this gives insight into what could happen for larger symbolic problems.

```
>  Norm( U^(-1).L^(-1).P^(-1).Yee
>           - IdentityMatrix(8,8), infinity );
```

$$0$$

2.2.6   Other Solvers

Solution of some Diophantine equations and some equations over finite fields is also possible in Maple. See ?isolve, ?msolve, and ?mod for details.

2.2.7   Systems of Polynomial Equations

Systems of polynomial equations in more than one variable are much more difficult to solve than univariate polynomials, in general. Numerical techniques, such as homotopy methods [45], are often (perhaps even usually) much more effective than the best symbolic techniques, those of "subresultants" and "Gröbner bases" [22]. Still, sometimes the symbolic techniques are what is wanted. Maple currently has implemented a heuristic substitution technique in its solve command, and a Gröbner basis method in its Groebner package.

```
>  restart;
>  eq1 := x^2 - y^2 - 1;
```

$$eq1 := x^2 - y^2 - 1$$

```
>  eq2 := x^3 - 3*x*y^2 + 3*x^2*y - y^3 + 1;
```

$$eq2 := x^3 - 3x\,y^2 + 3x^2\,y - y^3 + 1$$

By doing this session twice, I noticed that the following alias improves the presentation of the results.

```
>  alias( alpha=RootOf( 6*y+3+6*y^3+2*y^2, y ) );
```

$$\alpha$$

```
>  solve( {eq1, eq2}, {x,y} );
```

$$\{y = 0,\ x = -1\},\ \left\{x = -\frac{3}{7}\alpha + \frac{6}{7}\alpha^2 + \frac{5}{7},\ y = \alpha\right\}$$

The Gröbner basis technique produces an equivalent (in that it has the same roots) but more convenient set of equations to solve [19]. In fact, that is a loose but pragmatic definition of a Gröbner basis. The "ordering" chosen for the monomials involved helps to determine the special characteristics (and the speed of computation of) the Gröbner basis. We do not need to know details here, but rather how to select from the available orderings.

Choosing the purely lexicographic ordering `plex`, instead of the usually more economical total degree ordering `tdeg`, asks Maple to try to reduce the system of polynomials to a "triangular" system of polynomials. The system is "triangular" in that

1. the first equation is in only one variable,

2. the next contains only one new variable (usually linear in that new variable),

and so on. A `tdeg` ordering does not give a basis that has this property, in general, although we will see in the next section that it can still be used. Thus if one is able to compute a `plex`-ordered Gröbner basis, and if one has a reliable way to solve univariate polynomials, then one can solve the system written in terms of the Gröbner basis by solving a sequence of univariate polynomials. For this example, after solving the Gröbner basis we could see that it produced the same solutions as `solve` did.

```
>  gb := Groebner[gbasis]( {eq1,eq2}, plex(x,y) );
```

$$gb := \left[6\,y^2 + 3\,y + 6\,y^4 + 2\,y^3,\ 7\,x + 24\,y^3 + 7 + 27\,y + 2\,y^2\right]$$

```
>  factor( gb[1] );
```

$$y\,\left(6\,y + 3 + 6\,y^3 + 2\,y^2\right)$$

It is not clear which method is best for a given problem. The routine `solve`, when it works, is often faster. It can sometimes miss solutions, however. On the other hand, the computational complexity of computing `plex`-ordered Gröbner bases is very high, and moreover, the cheaper and stabler technique discussed below may well serve your purposes instead.

Remark. I remind you that the *numerical stability* of the resulting expressions (from `solve`) or the resulting system of polynomial equations (from `gbasis`) needs to be checked; just because one *can* make some particular mathematical transformation doesn't mean one *should* make that transformation. See the exercises.

A problem with free parameters.

The following problem is taken from [41]. It cannot be done by a `plex`-ordered Gröbner basis (the answer is just too complicated to be of any use even if we could calculate it in a reasonable length of time). But the approach here works in only a few seconds, with only a small amount of memory.

```
>   restart;
```

```
>   LSY[1] := x^3*y^2 + c1*x^3*y + y^2 + c2*x + c3;
```

$$LSY_1 := x^3 y^2 + c1 x^3 y + y^2 + c2 x + c3$$

```
>   LSY[2] := c4*x^4*y^2 - x^2*y + y + c5;
```

$$LSY_2 := c4 x^4 y^2 - x^2 y + y + c5$$

```
>   with( Groebner ):
```

The following step does not succeed if you ask for a `plex`-ordered basis, but with a total degree (`tdeg`) ordering it takes only a few seconds.

```
>   gb := gbasis( {LSY[1],LSY[2]}, tdeg(x,y) ):
```

```
>   ( ns, rv ) := SetBasis( gb, tdeg(x,y) ):
```

The first element returned from `SetBasis` is the normal set:

```
>   ns;
```

$$[1, \, y, \, x, \, y^2, \, x\,y, \, x^2, \, y^3, \, x\,y^2, \, x^2\,y, \, x^3]$$

The number of elements in the normal set tells us the number of complex zeros of the original system. The structure of the normal set is the structure of the eigenvectors of the matrices below. This will allow us to read off the roots of the original system from the eigenvectors. The routine `nops` counts the number of operands of a multiple object such as a set or list.

```
>   nops( ns );
```

$$10$$

The output of `MulMatrix` is a matrix representing multiplication by the variable it is given as an argument, in a certain commutative ring.

```
>   Mx := MulMatrix( x, ns, rv, gb, tdeg(x,y) ):
```

The entries of these sparse matrices are rational functions in the parameters c1 through c5. For example,

```
>  factor( Mx[10,3] );
```

$$
\begin{aligned}
(&c1\,c4\,c2^3 - c1^2\,c4^2\,c2\,c3^2 + 2\,c3^2\,c5\,c4^2\,c2\,c1 + c1^3\,c3\,c5 \\
&+ c1^2\,c4\,c3^2\,c5 + c3\,c2^2 + 2\,c1\,c4^3\,c2^3\,c3 + c4^2\,c3^2\,c2^2 \\
&+ c4^3\,c3^4\,c1 + 2\,c1^2\,c4^2\,c2^2\,c3 + 2\,c4\,c2^2\,c1^2\,c5 \\
&+ 2\,c3\,c2\,c1\,c5 + c4\,c1\,c3^3 + c4^3\,c1^3\,c3^3 + c4\,c1^3\,c2^2 \\
&+ 2\,c1^2\,c3\,c5^2)/(c1\,(c1^2\,c5 + c1\,c2 + c2^2\,c4) \\
&(c2 + c4^2\,c2\,c1^2 + c4^2\,c2\,c3 + c1\,c5))
\end{aligned}
$$

```
>  My := MulMatrix( y, ns, rv, gb, tdeg(x,y) ):
```

At this point, one may insert numerical values for c1 through c5 and find eigenvalues of a generic (convex random) combination of these two matrices, cluster any multiple roots, and use the Schur vectors to find the roots of the system. We see that there are generically 10 roots. We take random values for the parameters below, as an example, ignoring the possibility of multiple roots.

```
>  ( c1, c2, c3, c4, c5 ) := seq( rand()/1.0e12, i=1..5 ):
```

```
>  Digits := trunc(evalhf(Digits));
```
$$Digits := 14$$

```
>  with( LinearAlgebra ):
```

```
>  ( xvals, V ) := Eigenvectors( evalf( map(eval,Mx) ) ):
```

We may read off the corresponding y-values of the roots from the known structure of ns. Since the second element of ns is y, the second element of each eigenvector will be the y-value of the root (if each eigenvector is normalized so that its first entry is 1).

```
>  yvals := [ seq(V[2,i]/V[1,i],i=1..10) ]:
```

We substitute the computed values of x and y into the original equations, to see how nearly the computed quantities satisfy the original equations. To know how accurate our computed x and y are, we need more than just these residuals; we should look at how perturbations in these polynomials affect the roots. We do not do that here.

```
>  LSY[1], LSY[2];
```

$$x^3\,y^2 + .4855318020\,x^3\,y + y^2 + .2550506145\,x + .9529224743,$$
$$.6420653296\,x^4\,y^2 - x^2\,y + y + .1549126680$$

```
>    residuals := [ seq( eval([LSY[1],LSY[2]],
>         {x=xvals[i],y=yvals[i]}), i=1..10 ) ];
```

$$residuals := [[.8\,10^{-13} + 0.\,I, \;\; -.10\,10^{-13} + 0.\,I],$$
$$[.202\,10^{-12} - .21\,10^{-12}\,I, \;\; -.95\,10^{-12} + .20\,10^{-12}\,I],$$
$$[.202\,10^{-12} + .21\,10^{-12}\,I, \;\; -.95\,10^{-12} - .20\,10^{-12}\,I],$$
$$[-.3\,10^{-13} + 0.\,I, \;\; -.12\,10^{-12} + 0.\,I],$$
$$[.2\,10^{-13} - .1217\,10^{-11}\,I, \;\; -.24\,10^{-12} - .2\,10^{-12}\,I],$$
$$[.2\,10^{-13} + .1217\,10^{-11}\,I, \;\; -.24\,10^{-12} + .2\,10^{-12}\,I],$$
$$[.170\,10^{-11} - .126\,10^{-11}\,I, \;\; -.270\,10^{-11} + 0.\,I],$$
$$[.170\,10^{-11} + .126\,10^{-11}\,I, \;\; -.270\,10^{-11} + 0.\,I],$$
$$[-.72\,10^{-12} - .2852\,10^{-11}\,I, \;\; .24\,10^{-13} - .77\,10^{-12}\,I],$$
$$[-.72\,10^{-12} + .2852\,10^{-11}\,I, \;\; .24\,10^{-13} + .77\,10^{-12}\,I]]$$

Discontinuity of Gröbner bases

Gröbner bases are discontinuous. That is, if we have a set of polynomials that depends on a parameter, say $F(c) := \{f_i(x, y, z; c)\}$, then there may exist special points c^* for which the Gröbner basis is not what you get by first computing the Gröbner basis for $F(c)$ with a symbol for the parameter, and then putting $c = c^*$ in the result; instead, you have to compute the Gröbner basis for $F(c^*)$. In general, this requires the computation of a so-called *comprehensive Gröbner basis* [56]. This facility is not yet in Maple, and so we must find an alternative method. Here is one way to identify the special points.

```
>    restart;

>    with( Groebner ):

>    f[1] := c*x[1]^2*x[2] + 9*x[1]^2+2*x[1]*x[2]
>            +5*x[1] + x[2] - 3;
```

$$f_1 := c\,x_1{}^2\,x_2 + 9\,x_1{}^2 + 2\,x_1\,x_2 + 5\,x_1 + x_2 - 3$$

```
>    f[2] := 2*x[1]^3*x[2] + 6*x[1]^3
>            - 2*x[1]^2 - x[1]*x[2] - 3*x[1] -
>    x[2] + 3;
```

$$f_2 := 2\,x_1{}^3\,x_2 + 6\,x_1{}^3 - 2\,x_1{}^2 - x_1\,x_2 - 3\,x_1 - x_2 + 3$$

```
>    f[3] := x[1]^3*x[2] + 3*x[1]^3
>            + x[1]^2*x[2] + 2*x[1]^2;
```

$$f_3 := x_1{}^3\,x_2 + 3\,x_1{}^3 + x_1{}^2\,x_2 + 2\,x_1{}^2$$

In order to find the points c^*, you may use the following trick: Setting `infolevel[primpart] := 5` causes Maple to display all the equations that it removes as part of the computation as "content." These contents are assumed to be nonzero, and the resulting Gröbner basis is correct only if those assumptions hold, and the leading coefficients of the final basis are also nonzero. These equations thus define all the potential special points.

```
>  infolevel[primpart] := 5;
```

$$infolevel_{primpart} := 5$$

```
>  gb := gbasis( [ f[1], f[2], f[3] ], tdeg(x[1],x[2])):
```

```
Groebner/primpartscale:   remove content    1
Groebner/primpartscale:   total degree drops from   4    to    4
Groebner/primpartscale:   remove content    1
Groebner/primpartscale:   total degree drops from   4    to    4
Groebner/primpartscale:   remove content    1
Groebner/primpartscale:   total degree drops from   4    to    4
Groebner/primpartscale:   remove content    1
Groebner/primpartscale:   total degree drops from   3    to    3
Groebner/primpartscale:   remove content    1
Groebner/primpartscale:   total degree drops from   5    to    5
Groebner/primpartscale:   remove content    321110693270
Groebner/primpartscale:   total degree drops from   2    to    2
Groebner/primpartscale:   remove content    c
Groebner/primpartscale:   total degree drops from   4    to    3
Groebner/primpartscale:   remove content    343633073697
Groebner/primpartscale:   total degree drops from   3    to    3
Groebner/primpartscale:   remove content    c
Groebner/primpartscale:   total degree drops from   5    to    4
Groebner/primpartscale:   remove content    4498377794144977326691
Groebner/primpartscale:   total degree drops from   2    to    2
Groebner/primpartscale:   remove content    c^2
Groebner/primpartscale:   total degree drops from   5    to    3
Groebner/primpartscale:   remove content    10052256941514
Groebner/primpartscale:   total degree drops from   2    to    2
Groebner/primpartscale:   remove content    -18+6*c
Groebner/primpartscale:   total degree drops from   5    to    4
Groebner/primpartscale:   remove content    16729238502560864951 5|
Groebner/primpartscale:   total degree drops from   1    to    1
Groebner/primpartscale:   remove content    54-36*c+6*c^2
Groebner/primpartscale:   total degree drops from   5    to    3
Groebner/primpartscale:   remove content
8155509726840948638697169347909864132149 48684800
Groebner/primpartscale:   total degree drops from   1    to    1
Groebner/primpartscale:   remove content
47316150*c^2-9969050*c^3+771750*c^4+75557300-98253750*c
Groebner/primpartscale:   total degree drops from   5    to    1
Groebner/primpartscale:   remove content    252*c^2-1643*c+2966
Groebner/primpartscale:   total degree drops from   3    to    1
Groebner/primpartscale:   remove content    1
Groebner/primpartscale:   total degree drops from   1    to    1
Groebner/primpartscale:   remove content    1
Groebner/primpartscale:   total degree drops from   1    to    1
Groebner/primpartscale:   remove content    1
Groebner/primpartscale:   total degree drops from   1    to    1
```

That voluminous output must be searched for equations that define our special points c^*. If they are all nonzero, the following Gröbner basis is correct. So the only solution of these 3 equations in 2 unknowns for generic c is $x_1 = 0$, $x_2 = 3$.

```
>   gb;
```

$$[x_2 - 3, \; x_1]$$

But, choosing one of the equations defining special points,

```
>   54-36*c+6*c^2;
```

$$54 - 36\,c + 6\,c^2$$

```
>   factor( % );
```

$$6\,(-3 + c)^2$$

```
>   infolevel[primpart] := 0;
```

$$infolevel_{primpart} := 0$$

we get a very different Gröbner basis:

```
>   gb3 := gbasis( [ eval(f[1],c=3), f[2], f[3] ],
>           tdeg(x[1],x[2]));
```

$$gb3 := [-8\,x_1 - 5\,x_2 - 3 + 2\,x_2{}^2,$$
$$x_1\,x_2 + x_1 - x_2 + 3,$$
$$2\,x_1{}^2 - 3\,x_1 + 2\,x_2 - 6]$$

That is, for $c = 3$, there are many more solutions.

Exercises

1. Use the Turing factoring (LUDecomposition) to solve the following problem. For which values of k does the following augmented linear system have no, one, or infinitely many solutions?

$$\begin{bmatrix} k & k & 1 & k-1 \\ 1 & k & k & k^2-1 \\ k & 1 & k & k^3-1 \end{bmatrix} \qquad (2.1)$$

[Answer: Infinitely many if $k = 1$, none if $k = -\frac{1}{2}$, one solution otherwise.]

2. Solve $w^2 \exp w = x$ for w. See [11] for more information about the Lambert W function, which appears in the solution.

3. Choose several pairs of random bivariate polynomial systems of total degree 2 and try to solve them with solve and gsolve. Use time to time them. Which is better for problems of this size?

4. Find approximate solutions to the systems of the previous question using
   `fsolve`.

5. Find the solution of $x^2 y'' + xy' + y = \sin(x)$, $y(1) = 1$, $y'(1) = 0$.

6. Find the solution of $p_n = 2np_{n-1} + n$ with $p_0 = 1$, for all integers $n > 0$.

7. Five sailors and a monkey are stranded on a desert island. They go out picking coconuts all day, and agree to split the coconuts evenly in the morning.
   After everyone else is asleep, one of the sailors divides the coconut pile
   into five equal piles, and finds that there is one left over, which he (qui-
   etly) splits and gives to the monkey. He then hides one pile for himself, and
   puts the other four piles back together. Satisfied, he goes to sleep. Another
   sailor then awakens, divides the pile into five and finds there is one left over,
   which he gives to the monkey. He, too, hides his pile and puts the rest back
   together. Similarly, each sailor does the same in turn. In the morning, when
   all awaken, they—with guilty grins all around, except for the monkey—
   split the pile into five, finding that it divides evenly into five piles, with one
   left over for the monkey. What is the minimum possible (positive) number of coconuts they could have started with? [Answer: 15621 coconuts (it
   was a busy day, evidently). Note that -4 coconuts would work if we didn't
   restrict the answers to be positive!]

8. Suppose A is a symmetric n-by-n matrix with distinct eigenvalues. Then
   it is known that its eigenvalues are *perfectly conditioned*, just because of
   the symmetry of the matrix [24]. That is, small changes in the matrix will
   produce only small changes in the eigenvalues. Show by experiment that the
   Gröbner basis for the nonlinear system given by $Ax = \lambda x$ together with the
   normalization condition $\|x\|^2 = 1$ contains the characteristic polynomial
   of A (if the term ordering is taken with λ last). Since it is well known that
   most univariate polynomials are very poorly conditioned [58], conclude that
   computing a Gröbner basis can introduce spurious (and serious) numerical
   instability and thus may not be a practical way to calculate numerical roots
   of nonlinear systems.

9. Consider the example from [41] in Section 2.2.7. Find values of c_1, c_2,
   c_3, c_4, and c_5 for which the set of polynomials Maple computed as a total
   degree order Gröbner basis is, in fact, incorrect. As in the last section, set
   `infolevel[primpart] := 5;` before computing the Gröbner basis. This
   will cause certain polynomials, which Maple assumes to be nonzero, to be
   printed. Set some of these to zero, confounding Maple's assumptions, and
   redo the computation. Show that the resulting Gröbner basis is not the same
   as the one you get by taking the original Gröbner basis computed without

setting any combination of parameters to zero and taking the appropriate limit. Confirm thereby that Gröbner bases are not always continuous with respect to parameters.

2.3   Manipulations from Calculus

Calculus is the algebra of limits, derivatives, integrals, and series, at least as far as Maple is concerned. The corresponding commands in Maple are `limit`, `diff`, `int`, and `series`. These black boxes are at the same time stronger and weaker than a good human calculator. They are stronger because they have more "mathematical stamina" and can do longer calculations, and weaker because sometimes simple ad hoc techniques will give an answer where the standard techniques fail.

We begin with `diff`, the easiest of the routines to understand. There are several differentiation commands, including `diff`, `Diff`, and `D`. The command `Diff` is the "inert" form, and doesn't actually *do* anything; delaying evaluation using inert forms is sometimes useful (see Section 1.6.2). We will discuss `D` and other operators in a later chapter. See also the `linalg` routines `grad`, `jacobian`, and `hessian`.[5]

2.3.1   diff

The routine `diff` differentiates expressions representing functions. It can find the derivatives of all elementary functions and many special functions. You can extend it so that it knows how to differentiate your own functions.

```
>   restart;
```

```
>   F1 := diff( sin(x), x );
```
$$F1 := \cos(x)$$

```
>   F2 := diff( exp(sqrt(x+y)), x );
```
$$F2 := \frac{1}{2} \frac{e^{(\sqrt{x+y})}}{\sqrt{x+y}}$$

```
>   diff( F2, y );
```
$$-\frac{1}{4} \frac{e^{(\sqrt{x+y})}}{(x+y)^{(3/2)}} + \frac{\frac{1}{4} e^{(\sqrt{x+y})}}{x+y}$$

---

[5]The `linalg` package, as a linear algebra package, has been wholly superseded by the more efficient, more robust, and more programmable `LinearAlgebra` package. The differentiation routines from `linalg`, however, have not yet been moved to a "multivariate calculus" package, as they will be probably by the next version of Maple.

```
'diff/T' := proc( k, expr, x )
  local j, ans;
  if not type( k, 'integer' ) then
    'procname'( args )
  elif k<0 then
    diff( T(-k, expr ), x )
  elif k=0 then
    0
  elif k=1 then
    T( 0, expr )*diff( expr, x )
  else
    ans := -k*((-1)^(k-1)+1)/2 * T( 0, expr ) +
         2*k*add( T(k-1-2*j,expr), j=0..trunc((k-1)/2) );
    ans*diff( expr, x )
  end if;
end proc:
```

Figure 2.11: A Maple program to differentiate Chebyshev polynomials

```
>  Diff( sin(x), x ) = diff( sin(x), x );
```
$$\frac{\partial}{\partial x} \sin(x) = \cos(x)$$
```
>  value( lhs( % ) );
```
$$\cos(x)$$

See Section 1.6.2 for a discussion of the command value.

What follows is a somewhat complicated example of extending the knowledge of Maple's diff routine. The procedure in Figure 2.11 tells Maple how to differentiate Chebyshev polynomials $T_k(x)$, represented in Maple as T(k,x). Note that the chain rule for differentiation is incorporated in this routine itself.

For simpler examples of user-defined diff routines, see ?diff. The Chebyshev example in Figure 2.11 is not part of Maple already, and I felt it was useful to give a real but easily understood extension here. Note that if k is not a known integer on input, then the procedure returns "unevaluated."

```
>  restart;

>  read "D:/books/ess/programs/diffT.mpl";

>  diff( T(0,x^3), x );
```
$$0$$
```
>  diff( T(1,sin(x)), x );
```
$$T(0, \sin(x)) \cos(x)$$

```
>  diff( T(17,x), x );
```

$$17\,T(0,\,x) + 34\,T(16,\,x) + 34\,T(14,\,x)$$
$$+ 34\,T(12,\,x) + 34\,T(10,\,x) + 34\,T(8,\,x)$$
$$+ 34\,T(6,\,x) + 34\,T(4,\,x) + 34\,T(2,\,x)$$

```
>  eval( %, T=orthopoly[T] )
>           - diff( orthopoly[T](17,x), x ) ;
```

$$0$$

The above example showed some simple uses of `diff`, together with a nontrivial use of the user interface to `diff` to define the derivatives of the Chebyshev polynomials. See `?orthopoly` for a description of the orthogonal polynomial package, which contains routines for expansion of certain orthogonal polynomials. See also `?ChebyshevT`, which implements a generalization of the Chebyshev polynomials.

2.3.2   int

The next simplest command to use is `int`. This command will do both definite and indefinite integration of functions defined by expressions. The inert form `Int` provides a convenient interface to Maple's numerical quadrature routines. See Section 1.6.2 for more details.

```
>  restart;
```

```
>  int( sin(x), x );
```

$$-\cos(x)$$

Notice the conventional omission of the constant. If you want a constant, put it in yourself:

```
>  int( sin(x), x ) + C;
```

$$-\cos(x) + C$$

Maple can integrate functions that most first-year students can't.

```
>  int( ln(x)/(1+x), x );
```

$$\mathrm{dilog}(1 + x) + \ln(x)\ln(1 + x)$$

This integral is expressed in terms of the *dilogarithm function*. The dilogarithm function is

$$\mathrm{dilog}(x) = \int_1^x \frac{\ln t}{1 - t}\,dt\;.$$

This function often occurs in integrals arising from Feynman diagrams. Maple can do various computations with this function, including evaluating it for real values of x, plotting it, and computing series expansions.

Maple can also do *definite* integrals, as follows.

```
>  Digits := 30;
```

$$Digits := 30$$

```
>  Int( exp(x^2), x=0..1 ) = int( exp(x^2), x=0..1 );
```

$$\int_0^1 e^{(x^2)}\, dx = \frac{-1}{2}\, I\, \mathrm{erf}(I)\, \sqrt{\pi}$$

```
>  evalf( % );
```

$$1.46265174590718160880404858686$$
$$= 1.46265174590718160880404858686$$

The call to the inert function `Int` (note the capital letter `I` at the beginning) instead of `int`, followed by a call to `evalf`, invoked numerical integration. The analytical and numerical answers agreed. The numerical schemes used by Maple are sophisticated and powerful, involving analysis of singularities and an arbitrary-precision technique selected from several available.

Maple can evaluate integrals containing parameters. This is a powerful feature, that permits, for instance, Fourier series to be computed easily, as we saw in Section 1.1.3.

```
>  int( x^3*sin(m*x), x );
```

$$\frac{-m^3\, x^3 \cos(m\,x) + 3\, m^2\, x^2 \sin(m\,x) - 6 \sin(m\,x) + 6\, m\, x \cos(m\,x)}{m^4}$$

```
>  collect( %, [cos,sin], expand );
```

$$\left(-\frac{x^3}{m} + \frac{6\,x}{m^3}\right)\cos(m\,x) + \left(3\,\frac{x^2}{m^2} - \frac{6}{m^4}\right)\sin(m\,x)$$

Unevaluated integrals

'My Lord Morville,' replied Vandermast, 'it is altogether a cross matter and in itself disagreeing, that you should expect from me an answer to such a question.'
—E. R. Eddison, *A Fish Dinner in Memison*, Chapter XII.

If `int` returns unevaluated, it means one of the following things:

1. Maple has proved that the integral is not elementary.

2. Maple has given up on the integral.

3. Maple knows how to do the integral for some special case of the values of the parameters but is waiting for you to tell it whether the parameters fall in that class.

Here are some examples from each of the above categories.

```
>    restart;
```

```
>    int( tanh(x^2), x );
```

$$x + \int -2\,\frac{1}{\left(e^{(x^2)}\right)^2 + 1}\,dx$$

This means that the integral is not elementary. One can use `infolevel` to get the details of the proof from the Risch algorithm [22]. To prevent `int` from remembering its previous computation and thus giving us nothing interesting to trace, we tell `int` to forget what it has done.

```
>    forget( int );
```

```
>    infolevel[int] := 5;
```

$$infolevel_{int} := 5$$

```
>    int( tanh(x^2), x );
```

```
int/indef1:    first-stage indefinite integration
int/indef2:    second-stage indefinite integration
int/trighexp:    case of integrand containing exp and hyperbolic |
int/indef1:    first-stage indefinite integration
int/indef2:    second-stage indefinite integration
int/trighexp:    case of integrand containing exp and hyperbolic |
int/rischnorm:    enter Risch-Norman integrator
int/rischnorm:    exit Risch-Norman integrator
int/risch:    enter Risch integration
int/risch/algebraic1:    RootOfs should be algebraic numbers and
functions
int/risch:    the field extensions are
```

$$\left[x,\ e^{(x^2)}\right]$$

```
int/risch:    Introduce the namings:
```

$$\left\{\_th_1 = e^{(x^2)}\right\}$$

```
int/risch/int:    integrand is
```

$$\frac{\_th_1{}^2 - 1}{\_th_1{}^2 + 1}$$

```
int/risch/int:    integrand expressed as
```

$$1 - \frac{2}{\_th_1{}^2 + 1}$$

```
int/risch/ratpart:    integrating
```

$$-2\,\frac{1}{\_th_1{}^2 + 1}$$

```
int/risch/ratpart:    Hermite reduction yields
```

$$\int - 2\, \frac{1}{\_th_1{}^2 + 1}\, dx$$

```
int/risch/ratpart:
Rothstein's method - resultant is:
```

$$(1 - 2\,z\,x)^2$$

```
nonconstant coefficients: integral is not elementary
int/indef1:   first-stage indefinite integration
int/indef1:   first-stage indefinite integration
int/indef2:   second-stage indefinite integration
int/exp:   case of integrand containing exp
int/prpexp:   case ratpoly*exp(arg)
int/risch/exppoly:   integrating
```

$$1$$

```
int/risch/int:    integrand is
```

$$1$$

```
int/ratpoly/horowitz:   integrating
```

$$1$$

```
int/risch/ratpoly:    result is
```

$$x$$

```
int/risch/exppoly:    integral of the "constant term" is
```

$$x$$

```
int/risch:    exit Risch integration
int/indef1:   first-stage indefinite integration
int/indef1:   first-stage indefinite integration
int/indef2:   second-stage indefinite integration
int/exp:   case of integrand containing exp
int/prpexp:   case ratpoly*exp(arg)
int/rischnorm:    enter Risch-Norman integrator
int/rischnorm:    exit Risch-Norman integrator
```

$$x + \int - 2\, \frac{1}{\left(e^{(x^2)}\right)^2 + 1}\, dx$$

Those notes allow one to follow through with a proof that the integral is not elementary.

Here is an example from the second category, where Maple gives up.

```
>  restart;
>  f := sin( 3*arcsin(x) );
```

$$f := \sin(3\arcsin(x))$$

```
>   int( f, x );
```

$$\int \sin(3\arcsin(x))\,dx$$

As stated in Chapter 1, it can be shown that $\sin(3\arcsin(x))$ is a polynomial in x, namely $x(3-4x^2)$. Maple can do this integral if the proper conversions are done, and I expect that this weakness will be repaired by the next version of Maple. Moreover, if we ran this again with a high `infolevel`, we would see that Maple is not claiming to have proved that this integral is not elementary.

In case 3, some assumptions are needed before we can proceed.

```
>   restart;
```

```
>   int( t^n, t=0..1 );
```

$$\lim_{t\to 0+} -\frac{t^{(n+1)}-1}{n+1}$$

The unevaluated limit here tells the user that assumptions on n must be made before Maple can continue with the analysis. Sometimes an unevaluated integral means the same thing:

```
>   int( 1/(1+x^n), x=0..1 );
```

```
Definite integration: Can't determine if the integral is
convergent. Need to know the sign of --> n. Will now
try indefinite integration and then take limits.
```

$$\int_0^1 \frac{1}{1+x^n}\,dx$$

```
>   assume( n > 0 );
```

```
>   int( 1/(1+x^n), x=0..1 );
```

$$\int_0^1 \frac{1}{1+x^n}\,dx$$

with assumptions on n

That last seems to be a case from the second category, giving up too easily: Introducing a parameter allows Maple to continue!

```
>   int( 1/(a^n+t^n), t=0..1 );
```

$$\frac{\mathrm{hypergeom}\left(\left[1,\dfrac{1}{n}\right],\left[\dfrac{n+1}{n}\right],-a^{(-n)}\right)}{a^n}$$

with assumptions on n

Undefined integrals.

Maple may decide that your integral is undefined, and it will return undefined in that case.

```
>   restart;
>   int( 1/sin(x), x=-1..1 );
```
$$undefined$$

If you use the option CauchyPrincipalValue, you can force Maple to evaluate this integral.

```
>   int( 1/sin(x), x=-1..1, CauchyPrincipalValue );
```
$$0$$

See any complex variables text for a discussion of the Cauchy principal value. Sometimes the integral is infinite:

```
>   int( 1/x^2, x=-1..1 );
```
$$\infty$$

Continuity of antiderivatives

The following discusses a subtle bug in integration, present in most computer algebra systems. The bug is also present in many textbooks and tables so perhaps it is understandable why it persists. See [34] for a fuller discussion of the importance of being continuous.

```
>   restart;
>   f := 1/(2+sin(x));
```
$$f := \frac{1}{2 + \sin(x)}$$
```
>   plot( f, x=-3*Pi..3*Pi );
```

That plot can be found in Figure 2.12.

```
>   F := int( f, x );
```
$$F := \frac{2}{3} \sqrt{3} \arctan\left(\frac{1}{3} \left(2\tan\left(\frac{1}{2}x \right) + 1 \right) \sqrt{3} \right)$$

That integral is correct only in pieces. The function F is *discontinuous*, although there is a continuous antiderivative of f (as guaranteed by the fundamental theorem of calculus).

```
>   plot( F, x=-3*Pi..3*Pi, discont=true, colour=black );
```
This plot is shown in Figure 2.13, and we see clearly that this antiderivative is not satisfactory on any interval containing a discontinuity.

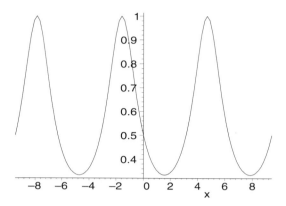

Figure 2.12: A continuous integrand that leads to a spuriously discontinuous antiderivative

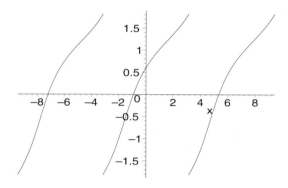

Figure 2.13: The spuriously discontinuous antiderivative of $1/(2 + \sin(x))$

```
>   limit( F, x=Pi, right );
```
$$-\frac{1}{3}\pi\sqrt{3}$$
```
>   limit( F, x=Pi, left );
```
$$\frac{1}{3}\pi\sqrt{3}$$

The function F has jump discontinuities at $x = (2n + 1)\pi$.

Maple uses some heuristics to correct this problem for a certain class of antiderivatives when it is asked to do a *definite* integral. The indefinite integral above is still unsatisfactory, though.

```
>  int( f, x=-2*Pi..2*Pi );
```

$$\frac{4}{3} \pi \sqrt{3}$$

```
>  evalf( % );
```

$$7.255197458$$

```
>  evalf( Int( f, x=-2*Pi..2*Pi ) );
```

$$7.255197457$$

Exercises

1. Find an antiderivative of $1/(2 + \sin x)$ that is continuous for all x. You may use the fact that $x/2 - \arctan(\tan(x/2))$ is piecewise constant. You may combine the arctangents by using the *rectifying transformation* $\tan^{-1}(t_1) + \tan^{-1}(t_2) \to \tan^{-1}((t_1+t_2)/(1-t_1t_2))$, which is itself true up to a piecewise constant.

2. Do the same for $1/(4 + \cos x)$.

3. If you have access to DERIVE, see whether it gives continuous antiderivatives for the exercises in this section.

4. (Contributed by Michael Wester) Get Maple to solve

$$\dot{x} = x \left(1 + \frac{\cos t}{2 + \sin t} \right) ,$$

$$\dot{y} = x - y ,$$

subject to the initial conditions $x(0) = 1$, $y(0) = 0$. Since these linear equations are infinitely differentiable, we do not expect any discontinuities in the solution. There are discontinuities in the solution from Maple 7. This bug may be corrected in a future version of Maple.

5. The time taken by a body to traverse part of its elliptical orbit around (say) the sun is proportional to the integral

$$\int_{\theta_0}^{\theta_1} r^2(\theta) \, d\theta ,$$

where the equation of the orbit is

$$r(\theta) = \frac{\text{constant}}{1 + \epsilon \cos \theta} .$$

Show that for Maple, time does not progress at the "stately pace of the planets" but instead does more of a "hip-hop."

2.3.3 limit

We have already seen examples of the use of `limit`, and no more really needs to be said about it. However, the following limit gets some people upset:

```
>  restart;
```

```
>  limit( sqrt(x), x=0 );
```

$$0$$

This is correct, because the square root function is well-defined (albeit complex) for negative x, and the limit is the same from all sides. There are other logical ways to arrive at this same conclusion even without using complex numbers, but some people choose to define limits by requiring both one-sided limits to exist and be the same; if that is the definition being used, then Maple's computation will not be what is desired.

2.3.4 series

Series in Maple are not just Taylor series. Some examples showing the possible mathematical forms follow. See also Section 3.3.

```
>  restart;
```

```
>  series( sin(x), x, 13 );
```

$$x - \frac{1}{6} x^3 + \frac{1}{120} x^5 - \frac{1}{5040} x^7 + \frac{1}{362880} x^9 - \frac{1}{39916800} x^{11} + O(x^{13})$$

```
>  series( x^x, x, 5 );
```

$$1 + \ln(x)\, x + \frac{1}{2} \ln(x)^2 x^2 + \frac{1}{6} \ln(x)^3 x^3 + \frac{1}{24} \ln(x)^4 x^4 + O(x^5)$$

The Maple routine `asympt` computes one-sided series about $x = \infty$.

```
>  asympt( 1/GAMMA(x), x, 2 );
```

$$\left(\frac{1}{2} \frac{\sqrt{2}}{\sqrt{\pi}\sqrt{\frac{1}{x}}} - \frac{1}{24} \frac{\sqrt{2}\sqrt{\frac{1}{x}}}{\sqrt{\pi}} + O\left(\left(\frac{1}{x}\right)^{(3/2)}\right) \right) \left(\frac{1}{x}\right)^x e^x$$

Puiseux series (series with fractional powers) are also part of the vocabulary:

```
>  series( sin(x)^(1/3), x );
```

$$x^{(1/3)} - \frac{1}{18}\, x^{(7/3)} - \frac{1}{3240}\, x^{(13/3)} + O\left(x^{(19/3)}\right)$$

This routine can do series computations for algebraic functions:

```
>  alias( alpha=RootOf(z^3+z+1, z) );
```

$$\alpha$$

```
>  series( RootOf(z^3+(1+x)*z+1, z), x );
```

$$\alpha + \left(\frac{2}{31}\alpha + \frac{9}{31}\alpha^2 + \frac{6}{31}\right) x$$

$$+ \left(\frac{25}{961}\alpha - \frac{27}{961}\alpha^2 - \frac{18}{961}\right) x^2$$

$$+ \left(\frac{67}{29791}\alpha - \frac{303}{29791}\alpha^2 - \frac{202}{29791}\right) x^3$$

$$+ \left(\frac{1732}{923521} + \frac{2598}{923521}\alpha^2 - \frac{4114}{923521}\alpha\right) x^4$$

$$+ \left(\frac{34208}{28629151} + \frac{51312}{28629151}\alpha^2 + \frac{1927}{28629151}\alpha\right) x^5 + O\left(x^6\right)$$

Reversion of series [29] is also possible in Maple. For example, consider the function defined by $t \tan t = x$.

```
>  x = t*tan(t);
```

$$x = t \tan(t)$$

```
>  map( series, %, t=Pi/4 );
```

$$x = \frac{1}{4}\pi + \left(\frac{1}{2}\pi + 1\right)\left(t - \frac{1}{4}\pi\right)$$

$$+ \left(\frac{1}{2}\pi + 2\right)\left(t - \frac{1}{4}\pi\right)^2 + \left(2 + \frac{2}{3}\pi\right)\left(t - \frac{1}{4}\pi\right)^3$$

$$+ \left(\frac{8}{3} + \frac{5}{6}\pi\right)\left(t - \frac{1}{4}\pi\right)^4 + \left(\frac{16}{15}\pi + \frac{10}{3}\right)\left(t - \frac{1}{4}\pi\right)^5$$

$$+ O\left(\left(t - \frac{1}{4}\pi\right)^6\right)$$

```
> solve( %, t );
```

$$\frac{1}{4}\pi + 2\frac{1}{\pi+2}\left(x - \frac{1}{4}\pi\right) - 4\frac{\pi+4}{(\pi+2)^3}\left(x - \frac{1}{4}\pi\right)^2$$
$$+ \frac{16}{3}\frac{36 + 14\pi + \pi^2}{(\pi+2)^5}\left(x - \frac{1}{4}\pi\right)^3 - \frac{64}{3}\frac{71\pi + 136 + 9\pi^2}{(\pi+2)^7}\left(x - \frac{1}{4}\pi\right)^4$$
$$- \frac{128}{15}\frac{-22\pi^3 - 5800 + 3\pi^4 - 3836\pi - 734\pi^2}{(\pi+2)^9}\left(x - \frac{1}{4}\pi\right)^5$$
$$+ O\left(\left(x - \frac{1}{4}\pi\right)^6\right)$$

Sometimes, one must increase the variable Order to get the desired result, especially if many leading terms in intermediate series cancel out:

```
> restart;
```

```
> f := ( sin(tan(x)) - tan(sin(x)) )/x^7;
```

$$f := \frac{\sin(\tan(x)) - \tan(\sin(x))}{x^7}$$

```
> series( f, x );
```

$$O(x^{-1})$$

```
> for k to 5 do
>     Order := Order + 1;
>     series( f , x );
> end do;
```

$$Order := 7$$
$$O(x^0)$$
$$Order := 8$$
$$-\frac{1}{30} + O(x)$$
$$Order := 9$$
$$-\frac{1}{30} + O(x^2)$$
$$Order := 10$$
$$-\frac{1}{30} - \frac{29}{756}x^2 + O(x^3)$$
$$Order := 11$$
$$-\frac{1}{30} - \frac{29}{756}x^2 + O(x^4)$$

The above examples show that Maple series can be a little different from Taylor series. In particular, care must be taken to get a precise definition of what Maple means by its use of the O-symbol.

Definition: An asymptotic sequence $\{\phi_k(x)\}$ is a sequence of functions defined near a point $x = a$ (a might be ∞) such that each element of the sequence is smaller than the preceding ones. We say that one function f is smaller than g if $f = o(g)$ near $x = a$; that is, $\lim_{x \to a} f(x)/g(x) = 0$.

One simple definition of the O (big-oh) symbol is that $f = O(g)$ if there exists a nonzero constant K such that $\lim_{x \to a} f(x)/g(x) = K$. For simplicity we ignore here the possibility of oscillations, which require more refined treatment. Thus f and g are roughly the same size (up to a multiplicative constant) near $x = a$. Maple allows the "constant" K to vary, but not by much: If Maple says that f is $O(x^6)$, it means that there is a function $k(x)$ such that $x = o(k(x))$ and $f(x) = O(k(x)x^6)$ in the standard sense, or else that $x = o(1/k(x))$ and $f(x) = O(k(x)x^6)$. For example, look at the series for x^x computed in the previous example, and note that the coefficients of the powers of x contain terms of the form $\log^n x$. For any n, these vary more slowly than x near the point of expansion, $x = 0$. See ?series and [6] for an alternative (but equivalent) description of the meaning of the O symbol in Maple.

When Maple computes a series, it ensures that each term is a member of an asymptotic sequence as above. The range of asymptotic sequences used by Maple is larger than most people normally use, and this explains the generalized series of Maple. However, there are many functions whose expansions require asymptotic sequences not in the lexicon of Maple. This is currently under development.

If you do not *want* such a generalized series, but rather want a strict Taylor series if it exists, then use the command `taylor` instead of `series`.

Exercises

1. Tell Maple about the function $s(x)$, whose derivative is $s'(x) = \tan(s(x))$, by defining a procedure `'diff/s'`. After that, ask Maple for the Taylor series of degree 3 for $s(x)$ about $x = 0$. This shows that `diff` and `series` are linked. Use `dsolve` to find an explicit expression for $s(x)$ and verify that your series is correct.

2. The *exponential generating function*, or "egf," for the Bernoulli numbers is $t/(\exp(t) - 1)$. That is, the Bernoulli numbers B_k, $k = 0, 1, 2, \ldots$, appear in the series expansion

$$\frac{t}{e^t - 1} = \sum_{k=0}^{\infty} \frac{B_k}{k!} t^k \ .$$

Compute the first eight nonzero B_k in this way, and compare with the built-in `bernoulli` function.

3. The *ordinary* generating function or "ogf," for the Catalan numbers is

$$F(x) = \frac{1 - \sqrt{1 - 4x}}{2x} = \sum_{k=0}^{\infty} f(k)x^k \,.$$

The difference between an ordinary generating function and an exponential generating function is the presence of $k!$ in the denominator of the terms of an exponential generating function. Compute the first eight Catalan numbers with `series`.

4. Compute the first eight of the indicated polynomials by using their ordinary generating functions given below. Take the series with respect to t.

(a) The Chebyshev polynomials of the first kind: $\text{ogf} = (1 - xt)/(1 - 2xt + t^2)$.

(b) The Chebyshev polynomials of the second kind: $\text{ogf} = 1/(1 - 2xt + t^2)$.

(c) The "tree polynomials" [39]: $\text{ogf} = 1/(1 + W(-t))^x$, where W is the Lambert W function.

2.4   Adding Terms versus the Finite-Difference Calculus

Most of the time in Maple programming, when one wants to write an expression containing the sum of several similar terms, one does *not* want to use the routine sum, although sum can be (ab)used to do it. The following examples illustrate the proper constructs to accomplish this common task. The routines are add, sum, and Sum. The corresponding constructs for multiplication are mul, product, and Product.

```
>   restart;
>   a := add( 1/k, k=1..10 );
```
$$a := \frac{7381}{2520}$$
```
>   k;
```
$$k$$
```
>   s := sum( 1/k, k=1..10 );
```
$$s := \frac{7381}{2520}$$
```
>   k;
```
$$k$$

```
>   S := Sum( 1/k, k=1..10 );
```

$$S := \sum_{k=1}^{10} \frac{1}{k}$$

```
>   evalf( S );
```

$$2.928968254$$

The routine `sum` is really the finite-difference analogue of the routine `int`, and should be used only if you want a symbolic antidifference (i.e., a sum of n terms expressed "in closed form").

```
>   add( 1/k, k=1..n );
```

```
Error, unable to execute add
```

That failed because n is a symbol, and add doesn't know how many terms to add. If we want a formula, we should use sum:

```
>   S[n] := sum( 1/k, k=1..n );
```

$$S_n := \Psi(n+1) + \gamma$$

```
>   limit( S[n] - log(n), n=infinity );
```

$$\gamma$$

The function $\Psi(x) = \Gamma'(x)/\Gamma(x)$ is the derivative of the log of the Γ function. The routine `sum` had to do a reasonable amount of work to find that out, and if $n = 10$ the work was wasted. This is not serious for this example, because evaluation takes too little time to measure in this case, but the advantage of simple addition over symbolic summation grows quickly with the complexity of the symbolic problem.

Like `Int`, there is also an inert form `Sum` that is merely a "placeholder" for the operation to be performed. The floating-point evaluation routine `evalf` knows about `Sum` and this provides a convenient user interface to floating-point evaluation of sums. However, `evalf/Sum` uses *Levin's u-transform* [57] to accelerate the convergence of a sum, and gives numerical values for convergent sums, slowly convergent sums, and even some divergent sums. Sometimes this is what is desired, as in the following example. If we define $c_1 = 1$ and

$$c_n = \frac{1}{1-n} \sum_{i=1}^{n-1} \binom{n-i+1}{i+1} c_{n-1},$$

where $\binom{n}{m} = n!/(m!(n-m)!)$ is the binomial coefficient, then the following program computes c_n for any n.

```
>    restart;
>    c := proc( n ) option remember;
>          -1/(n-1)*add(binomial(n-i+1,i+1)*c(n-i),i=1..n-1)
>       end proc:
>    c(1) := 1;
```

$$c(1) := 1$$

In Section 3.5 we will be discussing `option remember`. It makes the above procedure for the evaluation of a recurrence relation reasonably efficient. The manual assignment `c(1) := 1` places the base of the recursion into the remember table. This particular procedure was given to me by Bruno Salvy, more as a convenient way to e-mail me the recurrence relation than anything else.

```
>    seq(c(k),k=1..8);
```

$$1, -1, \frac{3}{2}, \frac{-8}{3}, \frac{31}{6}, \frac{-157}{15}, \frac{649}{30}, \frac{-9427}{210}$$

Note below the use of the inert Sum rather than sum to avoid needless symbolic processing here.

```
>    Bseries := v -> 'if'(v=0,1.,evalf(Sum('c(n)*v^(n-1)',
>    n=1..infinity)));
```

$$Bseries := v \rightarrow \text{'if'}\left(v = 0, \ 1., \ \text{evalf}\left(\sum_{n=1}^{\infty} \text{'}c(n)\, v^{(n-1)\text{'}}\right)\right)$$

We used the programmatic 'if' statement there. We had to check for the special case $v = 0$ because Maple substitutes $v = 0$ into the sum before it substitutes $n = 1, 2, 3, \ldots$, and then would simplify 0^{n-1} to zero.

Aside.   Maple uses the convention that $0^0 = 1$. While there exist relatively unimportant limits that can give results different than 1 as the base and exponent go to zero, this convention is useful in many contexts (for example, construction of Vandermonde matrices). For a good discussion of this convention, see [25].

```
>    Bseries(0);
```

$$1.$$

```
>    Bseries(0.001);
```

$$.9990014973$$

```
>    Bseries(0.05);
```

$$.9534459937$$

That series is divergent, but nevertheless the results from evalf/Sum are correct (the function $B(v)$ can be defined by a convergent infinite product, and the values computed by the acceleration method agree with the values computed by the product).

Remark. Levin's *u*-transformation is not a *regular* transformation, meaning that it can sometimes make convergent series divergent. Use `evalf( Sum( ... ) )` with caution.

Other methods for summing divergent series

One can program one's own sequence acceleration techniques. For example, in Figure 2.14 we find a short Maple program to evaluate sums by the Cesaro mean [27]:

$$\sum_{n=\ell}^{\infty} a_n := \lim_{m \to \infty} \frac{1}{m - \ell + 1} \sum_{k=\ell}^{m} a_k . \tag{2.2}$$

This procedure will give the sums

$$\sum_{n=0}^{\infty} (-1)^n = \frac{1}{2} , \tag{2.3}$$

$$\sum_{n=0}^{\infty} \cos(nx) = \frac{1}{2} , \tag{2.4}$$

$$\sum_{n=0}^{\infty} \sin(nx) = 0 . \tag{2.5}$$

It is an interesting exercise to plot the sum $\sum_{n=0}^{60} \cos(nx)$ on the interval $0 \le x \le \pi$, in point style, with say 1001 points. It becomes clear that the "average" value

```
CesaroSum := proc( a::algebraic, n::name, lower::integer)
   local avg, k, low, m, s ;
   low := 'if'( nargs<3, 0, lower );  # default lower limit is 0
   try
      # partial sums
      s    := sum( a, n=low..k );
      # averages of partial sums
      avg := sum( s, k=low..m )/(m+1-low);
      # limit of the averages of the partial sums
      return limit( avg, m=infinity );
   catch "numeric exception: division by zero":
      # Sometimes we can succeed if infinity is returned.
      NumericEventHandler( division_by_zero=((a,b,c)->infinity) );
      return CesaroSum( a, n, low );
   end try;
end proc:
```

Figure 2.14: A Maple program for Cesaro summation

of the sum is somewhere near $\frac{1}{2}$, but there is no possibility of the infinite sum being convergent in the ordinary sense.

Sum over RootOf.

Maple knows how to use symmetric functions to evaluate sums and products over the roots of polynomials. The following is a sum over all roots of a polynomial, done using rational means.

```
>    restart;
>    alias( alpha=RootOf(z^6+z+1, z) );
```

$$\alpha$$

```
>    Sum( 1/k, k=alpha );
```

$$\sum_{k=\alpha} \frac{1}{k}$$

```
>    value( % );
```

$$-1$$

Similarly, for products, .

```
>    Product( 1/(k+1), k=alpha );
```

$$\prod_{k=\alpha} \frac{1}{k+1}$$

```
>    value( % );
```

$$1$$

Exercises

1. Evaluate the following as "closed form" functions of n.

    (a) $\sum_{k=0}^{n} \sin(k\pi/n)$

    (b) $\sum_{k=0}^{n} k(k-7)$

    (c) $\sum_{k=1}^{n} 1/(k(k+1)^2)$

2. Evaluate the following correct to 5 decimal places using `evalf( Sum( ... ) )`.

    (a) $\sum_{k=0}^{\infty} 1/(k^2+1)$ [Answer: 2.0767]

    (b) $\sum_{k=0}^{\infty}(-1)^k k!/x^k$ for $x=10$. Note that the sum is divergent. Compare your answer to $x \int_{t=0}^{\infty} \exp(-t)/(x+t)\, dt$. Repeat for $x=100$. [Answer: 0.91563 and 0.99019]

    (c) $\sum_{k=1}^{\infty} k^{-1/3}$. Again the sum is divergent. Since each term in the sum is positive, does Maple's answer make any sense? See Section 1.6.2,

and the definition of the Riemann ζ-function (e.g., see ?Zeta).
[Answer: $\zeta(\frac{1}{3}) \approx -0.973$]

3. Use Maple to add the following sums, but do not use sum.

    (a) $\sum_{k=0}^{100}(-1)^k x^k/k!$ for $x = 30$, in exact rationals and then using $x = 30.0$ instead so the calculations are done in floating point arithmetic. Compare the answers. Repeat at higher settings of Digits, and explain the observed results. Note that the infinite sum is convergent (and indeed, the corresponding function $\exp(-x)$ is entire).

    (b) $\sum_{k=1}^{n} 1/k - \log(n + 1/2)$ for $n = 100, 1000$, and 10000. Compare your answers with γ, the Euler–Mascheroni constant (see ?gamma).

4. Evaluate $\prod_{k=\alpha} k/(2 - k^3)$ where α satisfies $1 + \alpha + \alpha^2 + \cdots + \alpha^n = 0$ for $n = 10, 20$, and 40. Use product and exact arithmetic.

2.5 Floating-Point Evaluation

Arbitrary-precision floating-point evaluation in computer algebra systems is over-rated for its utility (this is heresy; or, at least, a controversial opinion). The philosophy behind arbitrary precision is that you attempt to buy more accuracy in your answer by spending more time and memory on precision in your calculations. This works sometimes, for simple calculations, but is seldom required in real applications. On the other hand, it is intellectually satisfying, and is occasionally really needed, for special applications. Maple's evalf facility is a compromise between ease of programming and real efficiency for very large precisions, and is quite practical (more so, now, than when the first edition of this book was published, in part because machines are so much faster now).

There are two kinds of floating-point numbers used by Maple. The first is a Maple float, which is simply a pair of Maple integers wrapped in a call to the Float function: Float(i,j) means $i \cdot 10^j$, and prints in scientific notation. The second kind of floating point number used by Maple is a "hardware float," which is meant to match the characteristics of a floating-point number in actual hardware. Maple floats are used by the evalf routine, and hardware floats are used by evalhf.

The Maple routine evalf is robust and reliable (in a sense to be discussed below). It will call evalhf (see below) if it thinks it can safely do so (this isn't just a matter of the setting of Digits, because some chips do not evaluate some elementary functions to the full hardware-float accuracy). Thus at low settings of Digits you get some of the speed benefits of hardware floating-point. At higher settings of Digits the slower but more precise[6] software evalf subroutines come into

---

[6]A result is more precise if it has more digits; a result is more accurate if it has more correct digits.

play. If `Digits` is larger than `trunc(evalhf(Digits))`, which shows the Maple syntax for discovering the approximate number of decimal digits of a hardware float (the number of digits in a hardware float changes, of course, with the system you are using), then `evalhf` certainly cannot be used. Maple will sometimes also use other information such as conditioning of the problem in making its decision to use `evalhf` on a given problem.

On a single binary operation or evaluation of one single built-in function, Maple's `evalf` routine claims 0.6 ulp (units in the last place) *relative* accuracy. That is, the Maple result is the exact result, rounded correctly to the number of decimal digits requested by the user. The 0.6, rather than the theoretically attainable 0.5, represents a reasonable compromise between attainable accuracy and efficiency.

No claim is made for the accuracy of more than one operation. That is to say, no claim is made for the accuracy of evaluation of an arbitrary *expression*, and, indeed, any such claim would have to be backed up by interval arithmetic or intelligence about the conditioning of the particular expression being evaluated.

Maple knows about the evaluation of several special mathematical constants, π, e, γ, and some others. To avoid wasting cycles on the computation of π Maple stores π to 10,000 places, which means that the time taken to display 10,000 or fewer decimals of π is very small.[7] Of course, the time required to display 10,001 decimals of π is dramatically more! The other special constants are not stored to so many places (usually only 50 or so) since computing more digits of those constants is less frequently requested.

The following examples illustrate the use and limitations of `evalf`. We begin with simple evaluation of constants.

```
>   restart;
>   evalf( Pi, 50 );
        3.1415926535897932384626433832795028841971693993751
>   evalf( gamma, 50);
        .57721566490153286060651209008240243104215933593992
>   evalf( exp(1), 50 );
        2.7182818284590452353602874713526624977572470937000
>   evalf( Catalan, 50);
        .91596559417721901505460351493238411077414937428167
```

Now examine evaluation of expressions.

```
>   evalf( 1 + sqrt(2) + sin(Pi/6) + cos(Pi^5) + x/400 );
            2.632481588 + .002500000000 x
```

---

[7]The logic, of course, is that lots of students ask for lots of digits of π lots of times.

And now evaluation of special functions.

```
>  evalf( erf(5), 30);
```
$$.999999999998462540205571965150$$

```
>  Digits := 20;
```
$$Digits := 20$$

```
>  evalf( BesselJ(0,10) );
```
$$-.24593576445134833520$$

```
>  evalf( EllipticF(1/2,1/sqrt(2)) );
```
$$.53562273280540331970$$

```
>  evalf( GAMMA(0.2) );
```
$$4.5908437119988030532$$

Now a complex-valued example.

```
>  evalf( Zeta(1/2+30*I) );
```
$$-.12064228759004369991 - .58369121476370628876\, I$$

2.5.1 Using evalhf

The routine evalhf is necessary for faster plotting, especially for complicated three-dimensional plots. This results in speed gains of approximately a factor of 30 over the evalf routine. Probably another factor of 40 would bring it into line with the numerical speeds of a language such as C or FORTRAN. If you wish to use evalhf in your programs, I offer the following tips:

1. Concentrate everything in one evalhf-able routine. Bringing hardware floats into Maple (except in the form of hfarrays) forces an automatic conversion to Maple floats, which usually negates the advantage of using hardware floats in the first place. You *can* pass arrays of hardware floats around, from one subroutine to another. But any hardware floats that make it to the top level are converted.

2. Read the documentation carefully. Usually you will want to use evalhf to compute values in an array or vector, and you can do this; however, you will have to read and understand ?evalhf[arrays] and ?evalhf[var].

3. Don't nest recursion too deep: 15 levels seems to be the deepest you can go at the moment.

4. Avoid creating procedures for evalhf that have nested lexical scopes; these are not supported in evalhf.

5. Do not use more than ten formal parameters in the procedure which you pass to evalhf. If you need more parameters in your function, pack them into an array.

6. Do not use more than fifty local variables in a procedure that you pass to evalhf. This is usually a problem only with automatically generated procedures, such as are produced by optimize.

The following example shows the use of evalhf to evaluate the terms in the Taylor series of the solution of the differential equation

$$\dot{y}_1 = 2y_1(1 - y_2),$$
$$\dot{y}_2 = y_2(1 - y_1),$$

about a given point $t = a$, where $y_1(a) = y_1^{(0)}$ and $y_2(a) = y_2^{(0)}$. This is problem B1 from the DETEST problem suite [33]. [Incidentally, the exact solution for this problem can be expressed in terms of the Lambert W function [11]; see Section 2.2.3.] The code for the Taylor series, which appears in Figure 2.15, is easily derived from the Cauchy product formula, and requires $O(n^2)$ work to evaluate $O(n)$ terms in the Taylor series.

```
>  ya := array(1..2, [1., 3.]):

>  n := 200:

>  y := array( 1..2, 0..n ):

>  yf := array( 1..2, 0..n ):

>  st := time():
>      evalhf( ProblemB1(0., 0.1, ya, n, var(y)) ):
>  etime := time() -st;
```

$$etime := .214$$

```
ProblemB1 := proc(a, h, ya, n, y)
   local j,k,c,y1,y2;
   y[1,0] := ya[1];
   y[2,0] := ya[2];
   for k from 1 to n do
      c := 0;
      for j from 0 to k-1 do
         c := c + y[1,j]*y[2,k-j-1];
      end do;
      y[1,k] := 2*(y[1,k-1] - c)/k;
      y[2,k] := (y[2,k-1] - c)/k;
   end do;
end proc:
```

Figure 2.15: A Maple program written to use with evalhf

```
> Digits := trunc(evalhf(Digits))+1;
```
$$Digits := 15$$
```
> st := time():
>     evalf( ProblemB1(0., 0.1, ya, n, yf) ):
> etimef := time()-st;
```
$$etimef := 1.582$$
```
> etimef/etime;
```
$$7.39252336448598$$
```
> y[1,n];
```
$$.816462639698141420 \; 10^{61}$$
```
> yf[1,n];
```
$$.816462639698128 \; 10^{61}$$

One sees in this example that use of evalhf resulted in a speed increase of a factor of 7, roughly. Note the use of var to tell evalhf that the routine ProblemB1 is expected to return an array of values.

A slightly longer example

The program in Figure 2.16, namely a polyhedron-flake embedding of the 13-adic numbers (a generalization of an idea of Joachim von zur Gathen [55]), is used as follows.

```
> restart;
```

We set the path so this program can be found. On my system it is in the directory C:\local\mpl\adics; my system is a Windows machine and uses backslashes to separate directory names. As stated before, however, Maple was originally developed on Unix machines, and it expects paths to be separated with the forward slash /.

```
> currentdir( "C:/local/mpl/adics" );
> read "polyflake.mpl";
```

The parameter to polyflake (here 3) is the desired number of generations; the higher the number of generations the better the approximation to this three-dimensional fractal embedding of the 13-adic numbers (see [9] for details).

```
> bm := evalhf( polyflake(3) );
```

$$bm := \left[\, 1..28561 \; x \; 1..3 \; 2 - D \; Array \right.$$
$$Data \; Type : float[8] \quad Storage : rectangular$$
$$Order : C\_order \,]$$

We construct a plot by passing this hfarray (of 28,561 points) directly to the GUI by using the inert PLOT3D structure.

```
polyflake := proc( depth )
   local c, i, j, k, en, m, N, p, pts, r, s, t, xyz;
   # Icosahedral embedding hardwired (for now)
   N := 12;
   pts := array(1..N,1..3);
   s := 0.85065080835206;
   t := 0.52573111211912;
   # Coordinates of the vertices of an
   # icosahedron (from the geometry package)
   pts[ 1,2]:=  s; pts[ 1,3]:= t; pts[ 2,2]:=  s; pts[ 2,3]:= -t;
   pts[ 3,2]:= -s; pts[ 3,3]:= t; pts[ 4,2]:= -s; pts[ 4,3]:= -t;
   pts[ 5,1]:=  t; pts[ 5,3]:= s; pts[ 6,1]:=  t; pts[ 6,3]:= -s;
   pts[ 7,1]:= -t; pts[ 7,3]:= s; pts[ 8,1]:= -t; pts[ 8,3]:= -s;
   pts[ 9,1]:=  s; pts[ 9,2]:= t; pts[10,1]:=  s; pts[10,2]:= -t;
   pts[11,1]:= -s; pts[11,2]:= t; pts[12,1]:= -s; pts[12,2]:= -t;

   en := (N+1)^(depth+1); # total number of points
   xyz := array(1..en,1..3);

   # Initialize parents
   xyz[1,1] := 0.0;
   xyz[1,2] := 0.0;
   xyz[1,3] := 0.0;
   for i to N do
      xyz[1+i,1] := pts[i,1];
      xyz[1+i,2] := pts[i,2];
      xyz[1+i,3] := pts[i,3];
   end do;
   # depth level 0 done at this point.

   # Set up pointers
   p := 1+N;       # Final parent to be processed
   c := 2+N;       # Free position in stack for child
   r := 1.0/3.0;   # radius at current depth

   # Breadth-first traversal.
   for k to depth do
      # each parent has N children
      for j to p do
         for i to N do
            for m to 3 do
               xyz[c,m] := xyz[j,m]+r*pts[i,m];
            end do;
            c := c + 1;
         end do;
      end do;
      # Now all children & previous parents
      # become parents of smaller children
      r := r/3.0;
      p := c - 1;
   end do;
   # Return an hfarray suitable to pass to the GUI via PLOT3D.
   xyz
end proc:
```

Figure 2.16: A larger Maple program to use with evalhf

```
> bmp := PLOT3D( POINTS(bm) ):
```

This plot can be displayed in several orientations in an animation, as follows.

```
> b := n ->
> plots[display]( bmp, scaling=CONSTRAINED,
>       orientation=[n,60] );
> plots[display]( [seq(b(10*n),n=0..35)],
>       insequence=true );
```

We do not display that animation here. See

http://www.mapleapps.com/powertools/EssentialMaple7/EssentialMaple7.shtml

for an animated gif version, exported from Maple, or else create it yourself by executing the commands above.

2.5.2 Signed Zero

The numerics in Maple 6 and Maple 7 have been substantially improved over previous versions. In particular, in agreement with the IEEE854 standard for floating-point numbers, Maple now has a signed floating-point zero. More, it has a signed complex floating-point zero. This allows the range of certain functions (such as the logarithm) to be extended to allow symmetry over the floating-point complex numbers.

```
> ln( -1 + 0.*I );
```
$$0. + 3.14159265358979324\,I$$

```
> ln( -1 - 0.*I );
```
$$0. - 3.14159265358979324\,I$$

This signed complex zero follows the usual rules of arithmetic, and allows certain programs (such as the divide-and-conquer eigenvalue computation method described in [37]) to be written more concisely and efficiently.

Further, Maple now allows the user to control the rounding mode and to use special-purpose event handlers for numeric events. See ?numerics and ?NumericEventHandler for details. An example of the use of NumericEvent-Handler can be found in Figure 2.14.

2.6 The Most Helpful Maple Utilities

My candidate for the most useful Maple command of all (aside, of course, from help or ?) is the read statement. Without this to read in files of Maple commands, Maple would be almost unusable. For Maple programs longer than about four lines, always use an editor to create files of Maple commands and the read statement to read them in. As mentioned in Chapter 1, Waterloo Maple Inc. recommends the vim editor, freely available from www.vim.org. I personally prefer WinEdt (www.winedt.com, and note that this is not WinEdit) on Windows systems and emacs on Unix systems. Using an editor for programming is just common sense.

2.6.1   I/O Utilities

There are two other I/O utilities in Maple, `sscanf` and `printf`, based on the C utilities of the same names. These, demonstrated below, are useful for formatted I/O. See also `readdata`.

```
>   for i to 10 do
>       printf( "i = %+2d and i^(1/2) = %+6.3f",
>               i, evalf(sqrt(i)) );
>   end do;
```

```
i = +1 and i^(1/2) = +1.000i = +2 and i^(1/2) = +1.414i = +3|
```

I forgot the newline character \n in the format string. This is a common error. I truncated the line above so it would fit on the page. We now try again:

```
>   for i to 10 do
>       printf( "i = %+2d and i^(1/2) = %+6.3f\n",
>               i, evalf(sqrt(i)) );
>   end do;
```

```
i = +1 and i^(1/2) = +1.000
i = +2 and i^(1/2) = +1.414
i = +3 and i^(1/2) = +1.732
i = +4 and i^(1/2) = +2.000
i = +5 and i^(1/2) = +2.236
i = +6 and i^(1/2) = +2.449
i = +7 and i^(1/2) = +2.646
i = +8 and i^(1/2) = +2.828
i = +9 and i^(1/2) = +3.000
i = +10 and i^(1/2) = +3.162
```

The +2d in the format string means a signed decimal integer in a field of width 2. Since +10 takes 3 characters it automatically widens the format (unlike FORTRAN which would have printed **). Let's try it with 3.

```
>   for i to 10 do
>       printf( "i = %+3d and i^(1/2) = %+6.3f\n",
>               i, evalf(sqrt(i)) );
>   end do;
```

```
i =  +1 and i^(1/2) = +1.000
i =  +2 and i^(1/2) = +1.414
i =  +3 and i^(1/2) = +1.732
i =  +4 and i^(1/2) = +2.000
i =  +5 and i^(1/2) = +2.236
i =  +6 and i^(1/2) = +2.449
i =  +7 and i^(1/2) = +2.646
i =  +8 and i^(1/2) = +2.828
i =  +9 and i^(1/2) = +3.000
i = +10 and i^(1/2) = +3.162
```

```
> for i to 10 do
>    printf( "i = % 3d and i^(1/2) = % 6.3f\n",
>           i, evalf(sqrt(i)) );
> end do;
```

```
i =   1 and i^(1/2) =  1.000
i =   2 and i^(1/2) =  1.414
i =   3 and i^(1/2) =  1.732
i =   4 and i^(1/2) =  2.000
i =   5 and i^(1/2) =  2.236
i =   6 and i^(1/2) =  2.449
i =   7 and i^(1/2) =  2.646
i =   8 and i^(1/2) =  2.828
i =   9 and i^(1/2) =  3.000
i =  10 and i^(1/2) =  3.162
```

A blank instead of a + before the 3d in the format string means that positive numbers are prefixed by a blank and negative numbers by a − sign.

2.6.2 `alias` **and** `macro`

The next most useful Maple commands are `alias` and `macro`. These allow you to use short names for convenience when the "real" name of the object is quite long and awkward to type or read. I use `alias` most often together with `RootOf`, but it is also helpful in programming when you want to use long names to avoid conflicts with global variables but wish to use short names when typing. Note that it has been used throughout this chapter, and in particular in the last example of section 2.4.

2.6.3 Interacting with the Operating System and External Calls

The `ExternalCalling` module provides an interface to the `define_external` command. This feature is new to Maple 7 and I have not used it yet.

See also `?system` for details on the old way to call other programs from within Maple.

2.6.4 Mapping Functions Onto Compound Objects

The next two most useful Maple commands are arguably `map` and `unapply`. The first maps a function onto each element of some object (list, array, or set), and the second we have already seen used to create an operator from an expression. Similarly useful commands are `zip` (nothing to do with file compression, it just zips lists together like a zipper) and `select`; advanced functional programmers may wish to use `foldl`.

```
>   restart;
```

```
>   with( LinearAlgebra ):
```

```
>   A := Matrix( 3, 3, (i,j)->1/(i+j+2) );
```

$$A := \begin{bmatrix} \dfrac{1}{4} & \dfrac{1}{5} & \dfrac{1}{6} \\[2mm] \dfrac{1}{5} & \dfrac{1}{6} & \dfrac{1}{7} \\[2mm] \dfrac{1}{6} & \dfrac{1}{7} & \dfrac{1}{8} \end{bmatrix}$$

```
>   map( t->1/t, A );
```

$$\begin{bmatrix} 4 & 5 & 6 \\ 5 & 6 & 7 \\ 6 & 7 & 8 \end{bmatrix}$$

The command Map does the same thing, but *in place* and so overwrites the original matrix A. This is an important consideration for large matrices.

```
>   s := [ seq(i^3, i=0..5) ];
```

$$s := [0, 1, 8, 27, 64, 125]$$

```
>   r := map( m->sin(Pi*m/64), s );
```

$$r := \left[0, \sin\left(\frac{1}{64}\pi\right), \sin\left(\frac{1}{8}\pi\right), \sin\left(\frac{27}{64}\pi\right), 0, -\sin\left(\frac{3}{64}\pi\right) \right]$$

```
>   pts := zip( (a,b)->[a,b], s, r );
```

$$pts := \left[[0, 0], \left[1, \sin\left(\frac{1}{64}\pi\right)\right], \left[8, \sin\left(\frac{1}{8}\pi\right)\right], \right.$$
$$\left. \left[27, \sin\left(\frac{27}{64}\pi\right)\right], [64, 0], \left[125, -\sin\left(\frac{3}{64}\pi\right)\right] \right]$$

The select command picks out members of a multiple object that satisfy a given criterion, in this case that the element has somewhere in it the number "27."

```
>   select( has, pts, 27 );
```

$$\left[\left[27, \sin\left(\frac{27}{64}\pi\right) \right] \right]$$

The `remove` command is the opposite of `select`.

```
>   remove( has, pts, 0 );
```

$$\left[\left[1, \sin\left(\frac{1}{64}\pi\right)\right], \left[8, \sin\left(\frac{1}{8}\pi\right)\right],\right.$$
$$\left.\left[27, \sin\left(\frac{27}{64}\pi\right)\right], \left[125, -\sin\left(\frac{3}{64}\pi\right)\right]\right]$$

You could do both those at once (more efficiently) using `selectremove`.

2.6.5 Code Generation

We look now at the routines from the code-generation package `codegen`, namely `makeproc`, `fortran` and `C`. They convert Maple expressions, vectors, or matrices into Maple, FORTRAN or C code. The code optimization features, which use a "janitorial" approach of trying to clean up an existing messy expression, are sometimes very useful.

```
>   restart;
```

```
>   with( LinearAlgebra ):
```

```
>   with( codegen ):
```

```
Warning, the protected name MathML has been redefined and unpro|
```

```
>   A := Matrix( 3, 3, (i,j)->1/(i+j+t) );
```

$$A := \begin{bmatrix} \dfrac{1}{2+t} & \dfrac{1}{3+t} & \dfrac{1}{4+t} \\[2mm] \dfrac{1}{3+t} & \dfrac{1}{4+t} & \dfrac{1}{5+t} \\[2mm] \dfrac{1}{4+t} & \dfrac{1}{5+t} & \dfrac{1}{6+t} \end{bmatrix}$$

```
>   p := CharacteristicPolynomial( A, x );
```

$$p := (-4 - 96\,x^2\,t^7 - 3\,x^2\,t^8 + 5208\,x^3\,t^6 + 570\,x^3\,t^7 + 36\,x^3\,t^8 + x^3\,t^9$$
$$+ 120\,x\,t^4 + 6\,x\,t^5 + 6288\,x - 158400\,x^2 - 334560\,x^2\,t$$
$$- 306716\,x^2\,t^2 + 3864\,x\,t^2 + 962\,x\,t^3 + 7780\,x\,t$$
$$- 159408\,x^2\,t^3 - 51371\,x^2\,t^4 + 429120\,x^3\,t + 466512\,x^3\,t^2$$
$$+ 291740\,x^3\,t^3 + 115764\,x^3\,t^4 + 30249\,x^3\,t^5 + 172800\,x^3$$
$$- 10512\,x^2\,t^5 - 1334\,x^2\,t^6)/((2+t)\,(4+t)^3\,(6+t)\,(5+t)^2$$
$$(3+t)^2)$$

```
>  fortran( p, optimized );

       t1 = t**2
       t4 = x**2
       t5 = t4*x
       t6 = t1**2
       t7 = t6**2
       t12 = t1*t
       t13 = t6*t12
       t32 = -4+3864*x*t1+t5*t7*t+36*t5*t7-96*t4*t13-3*t4*t7-|
      #962*x*t12-159408*t4*t12-51371*t4*t6+466512*t5*t1+42912|
      #0*t5*t12
       t35 = t6*t
       t47 = t6*t1
       t58 =115764*t5*t6+30249*t5*t35+570*t5*t13+6*x*t35-1051|
      #88*x+120*x*t6+5208*t5*t47+7780*x*t-306716*t4*t1+172800|
      #4-1334*t4*t47
       t63 = 4+t
       t64 = t63**2
       t71 = (5+t)**2
       t75 = (3+t)**2
       t78 = (t32+t58)/(2+t)/t64/t63/(6+t)/t71/t75
```

We could send the above output to a file using the `filename` option. However, perhaps we should first make it a little more efficient, as follows.

```
>  L1 := [ optimize(p) ];
```

$L1 := [t1 = x^2,\ t2 = t^2,\ t3 = t2^2,\ t4 = t3^2,\ t7 = t1\,x,\ t14 = t3\,t,$

$t21 = t2\,t,\ t32 = -4 - 3\,t1\,t4 + 36\,t7\,t4 + t7\,t4\,t + 120\,x\,t3$
$\qquad + 6\,x\,t14 - 334560\,t1\,t + 3864\,x\,t2 + 962\,x\,t21 + 7780\,x\,t$
$\qquad - 159408\,t1\,t21 - 51371\,t1\,t3 + 429120\,t7\,t,\ t43 = t3\,t2,$

$t48 = t3\,t21,\ t58 = 466512\,t7\,t2 + 291740\,t7\,t21$
$\qquad + 115764\,t7\,t3 + 30249\,t7\,t14 - 10512\,t1\,t14 - 1334\,t1\,t43$
$\qquad - 306716\,t1\,t2 + 570\,t7\,t48 + 172800\,t7 + 5208\,t7\,t43$
$\qquad - 96\,t1\,t48 - 158400\,t1 + 6288\,x,\ t63 = 4 + t,\ t64 = t63^2,$

$t71 = (5 + t)^2,\ t75 = (3 + t)^2,$

$$t78 = \frac{t32 + t58}{(2 + t)\,t64\,t63\,(6 + t)\,t71\,t75}]$$

```
>  cost( L1 );
```

$59\,multiplications + 16\,assignments + 30\,additions + 6\,divisions$

```
> L2 := [ optimize(p,tryhard) ];
```

$$L2 := [t22 = 4 + t,\ t11 = t^2,\ t15 = t\,t11,\ t18 = t15^2,\ t16 = t11^2,$$
$$t12 = t22^2,\ t10 = t\,t18,\ t6 = t\,t16,\ t2 = 3 + t,\ t1 = 5 + t,\ t3 = ($$
$$-4 + (962\,t15 + 120\,t16 + 6\,t6 + 3864\,t11 + 6288 + 7780\,t$$
$$+ (-158400 - 306716\,t11 - 159408\,t15 - 334560\,t$$
$$- 1334\,t18 - 96\,t10 - 10512\,t6 + (-51371 - 3\,t16)\,t16 + ($$
$$429120\,t + 5208\,t18 + 172800 + 570\,t10 + 30249\,t6$$
$$+ 466512\,t11 + 291740\,t15 + (115764 + (t + 36)\,t16)\,t16)$$
$$x)x)x)/((2 + t)\,t22\,t12\,(6 + t)\,t1^2\,t2^2)]$$

```
> cost( L2 );
```

$$30\,additions + 11\,assignments + 33\,multiplications + 6\,divisions$$

That is somewhat better. Note that the cost of using the `tryhard` option can be dramatically more than the default option, on large examples, but if improved execution speed of the optimized formula is important, it can be worth it.

```
> C(L2);
```

```
        t54 = 4.0+t;
        t42 = t*t;
        t47 = t*t42;
        t50 = t47*t47;
        t51 = t50*t;
        t48 = t42*t42;
        t45 = t54*t54;
        t41 = x*x;
        t40 = t48*t48;
        t35 = t47*t42;
        t34 = 3.0+t;
        t33 = 5.0+t;
        t1 = (-4.0+(-10512.0*t35-3.0*t40-334560.0*t-1334.0*t50-5137|
   -159408.0*t47-96.0*t51-306716.0*t42-158400.0)*t41+(6288.0+7780.0*|
   *t35+962.0*t47+120.0*t48+(172800.0+5208.0*t50+30249.0*t35+36.0*t4|
   1+429120.0*t+115764.0*t48+291740.0*t47)*t41+(3864.0+(t51+466512.0|
   t42)*x)/(2.0+t)/t54/t45/(6.0+t)/(t33*t33)/(t34*t34);
```

As usual, I truncated that output so it would fit on the page.

Exercises

1. Read the help file entries for `sscanf` and `readdata` and try them out.

2. Write a C or FORTRAN program that uses the cubic formula to find the roots of a given cubic equation. Use Maple to generate the cubic formula, and C or `fortran` to write the fragment of code at the heart of your program. Now worry about numerical stability.

3. Use the `define_external` command to call some program external to Maple, perhaps a numerical program for solving a boundary value problem. Use the results in Maple.

2.7   Plotting in Maple

In that perfect hour all shadows had left earth and sky, and but form and colour remained: form, as a differing of colour from colour, rather than as a matter of line and edge (which indeed were departed with the shadows)...

—E. R. Eddison, *A Fish Dinner in Memison*, Chapter V.

Maple is not primarily a visualization language—it will not provide publication-quality graphics for you. However, its plotting facilities are improving (Release 7 plots are much better than those of previous versions) and are powerful enough to provide much insight; and its plots can be improved by draftists[8] for publication if necessary. "Raw" Maple plots are used in this book, however, since fidelity is more important in this context.

All the plots printed here are inferior to what you get by issuing the Maple commands and viewing the plots "live." In particular, three-dimensional plots can be rotated by dragging them with a mouse; the colours can be altered (all plots are printed here in black and white); and animations are possible. Several animated plots produced by Maple can be seen at my web site,

http://www.apmaths.uwo.ca/~rcorless ,

or at the Maple site

http://www.maplesoft.com .

2.7.1   Two-Dimensional Plots

Maple has facilities for plotting graphs of real functions represented by expressions, operators, or data; it can plot functions represented in Cartesian coordinates, polar coordinates, or parametrically. The relevant Maple command is `plot`.

As a simple example, consider using Maple to plot some real values of the Riemann ζ-function.

```
>  restart;
>  plot( Zeta(t), t=-3..3, y=-3..3,
>          discont=true, colour=black );
```

---

[8]A *draftist* is a person skilled in the techniques of drafting; what used to be called a draughtsman, in other words. It is quite interesting that for mathematical plots, real human people are still better in some ways for producing intelligible and good-looking results.

This graph was printed to an encapsulated PostScript file, and is shown in Figure 2.17. The option `discont=true` was used as a signal to Maple that the expression was discontinuous somewhere in the plot, and Maple then tried not to draw a line from the top to the bottom of the plot at the singularity (successfully in this case).

We can also put more than one plot on a graph, as follows.

```
>   alias( W=LambertW );
```

$$W$$

```
>   plot( {W(x),W(-1,x)}, x=-0.5..1.5, y=-4..1,
>            colour=black );
```

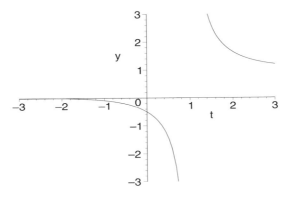

Figure 2.17: The Riemann ζ-function

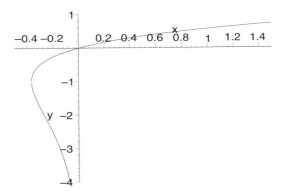

Figure 2.18: The Lambert W function, drawn directly

This plot, shown in Figure 2.18, shows the graph of the two real branches of $W(x)$ (see [11]) plotted on the same curve. There is some difficulty with plotting close to the branch point $(x, y) = (-1/e, -1)$, which shows up here as a "hole" in the graph. Alternatively, we can plot this function parametrically, as follows.

```
>   plot( [t*exp(t), t, t=-4..1], view=[-0.5..1.5,-4..1],
>           colour=black );
```

This parametric plot uses the definition of $W(x)$ as the number w such that $w \exp w = x$ to plot the graph completely (it is also much faster than the previous plot, because $\exp x$ is a "more built-in" function). This graph is shown in Figure 2.19.

Now let us plot a collection of functions; say, some of the Chebyshev polynomials.

```
>   plot( {seq(orthopoly[T](k,x),k=0..20)},
>           x=-1..1, y=-1..1,
>           axes=BOXED, colour=black,
>           scaling=CONSTRAINED, numpoints=101 );
```

This graph is shown in Figure 2.20. Note the interesting curves that appear to be suggested by the places where the Chebyshev polynomials do *not* go. See [48] for a detailed explanation of these curves. The loci of these "negative" curves are given by the equation

$$T_2(y) = T_q(x)$$

for $q = 1, 2, 3, \ldots$, and this can be solved parametrically by $y = \pm T_q(t)$, $x = T_2(t)$, because of the remarkable *composition property* of the Chebyshev

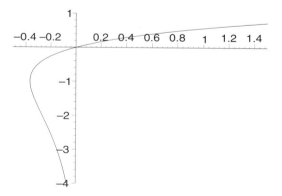

Figure 2.19: The Lambert W function, drawn parametrically

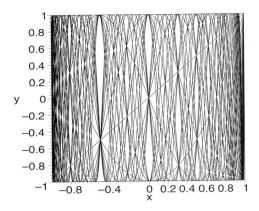

Figure 2.20: The first twenty Chebyshev polynomials

polynomials, $T_m(T_n(x)) = T_n(T_m(x)) = T_{mn}(x)$ (see [48]). We plot the first few of these curves as follows.

```
>   with( orthopoly ):
>   plot( {seq( [T(2,t),T(k,t),t=-1..1], k=1..5) ,
>          seq( [T(2,t),-T(k,t),t=-1..1],k=1..5)},
>          view=[-1..1,-1..1], colour=black,
>          scaling=CONSTRAINED, axes=BOXED );
```

This graph is shown in Figure 2.21.

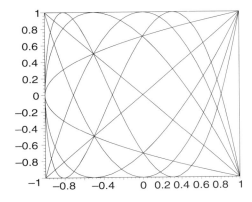

Figure 2.21: The intersection loci

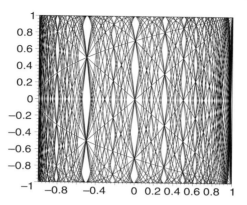

Figure 2.22: More intersection loci

I cannot resist looking at the plot generated by the first twenty of these curves:

```
>    plot( {seq( [T(2,t),T(k,t),t=-1..1], k=1..20) ,
>              seq( [T(2,t),-T(k,t),t=-1..1],k=1..20)},
>              view=[-1..1,-1..1], colour=black,
>              scaling=CONSTRAINED, axes=BOXED );
```

These curves, plotted in Figure 2.22, suggest that a similar analysis to that for the Chebyshev polynomials is possible here. See [12] for such an analysis.

We may use the `plot` command indirectly, as with the following use of the student package.

```
>    with(student):
>    leftbox(1/(1+t), t=0..1, 6, colour=black);
```

This plot is shown in Figure 2.23.

Now let us explore graphing partial sums of Fourier series. Compare this section with the sample session in Section 1.1.3.

Consider this simple periodic function.

```
>    restart;
>    f := 1/(2+sin(theta));
```

$$f := \frac{1}{2 + \sin(\theta)}$$

That function is periodic, and neither even nor odd. Hence its Fourier series will contain both cosine and sine terms. We use the standard integrals to compute a and b.

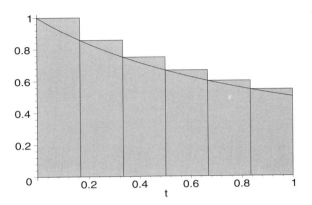

Figure 2.23: Riemann sums for $1/(1+t)$

```
>  a := n -> Int( f*cos(n*theta), theta=0..2*Pi )/Pi;
```

$$a := n \to \frac{\displaystyle\int_0^{2\pi} f \cos(n\,\theta)\,d\theta}{\pi}$$

```
>  b := n -> Int( f*sin(n*theta), theta=0..2*Pi )/Pi;
```

$$b := n \to \frac{\displaystyle\int_0^{2\pi} f \sin(n\,\theta)\,d\theta}{\pi}$$

```
>  Approx := N -> add( b(k)*sin(k*theta), k=1..N )
>          +add( a(k)*cos(k*theta), k=1..N ) + a(0)/2;
```

$$Approx := N \to \operatorname{add}(b(k)\sin(k\,\theta),\, k = 1..N)$$
$$+ \operatorname{add}(a(k)\cos(k\,\theta),\, k = 1..N) + \frac{1}{2}\,a(0)$$

```
>  err := evalf( f - Approx(5) ):
>  plot( err, theta=0..2*Pi );
```

The plot is shown in Figure 2.24, and shows that the Fourier series gives a good representation of the function.

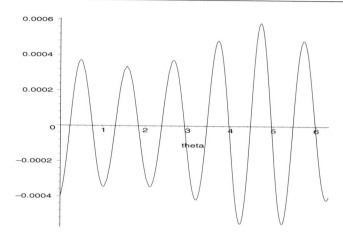

Figure 2.24: The error in representing f by the first five terms in its Fourier series

A highly discontinuous function

We now consider trying to graph the Gauss map $G : t \rightarrow t^{-1} \mathrm{mod}\ 1$, from the theory of continued fractions [7]. This function has discontinuities at $t = 1/n$, for all positive integers n. In Maple, this function can be defined as follows.

```
>   restart;
```

```
>   G := t -> frac(1/t);
```

$$G := t \rightarrow \mathrm{frac}(\frac{1}{t})$$

```
>   plot( G, 0..1, 0..1, numpoints=101, colour=black );
```

This plot is shown in Figure 2.25. It is a very ugly plot. Now, perhaps that was unfair; there is a serious singularity at the origin, and we will have to deal with that ourselves.

What follows is an extended Maple session, which is intended for you to follow as if you were looking over my shoulder as I type. A significant amount of my own Maple knowledge was acquired by actually looking over the shoulders of various Maple experts. It's a good method. I will try to anticipate all your questions, but you can always try ? on any mysterious construct.

The session will provide an example of how to plot point data in Maple, and more than one such plot on a graph. Let us choose 11 (almost) equally spaced points, in y, ranging from $1 - 10^{3-\mathrm{Digits}}$ down to 0 (such a strange number was chosen only after some fiddling).

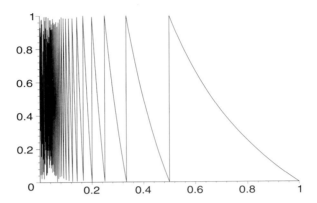

Figure 2.25: First attempt to plot the Gauss map $G(t)$

```
>   y := [ seq( k/10., k=0..10 ) ];
```

$$y := [0., .1000000000, .2000000000,$$
$$.3000000000, .4000000000, .5000000000,$$
$$.6000000000, .7000000000, .8000000000,$$
$$.9000000000, 1.000000000]$$

```
>   y[11] := y[11] - Float(1,3-Digits);
```

$$y_{11} := .999999900$$

This replaces the eleventh element in y with the desired quantity. Now let us specify the t-values that correspond to those y-values. There should be infinitely many, one on each piece of the graph, but we will make do with 100 pieces; which gives 1100 t-values.

```
>   t := Vector[row]( 1..1100 ):
>   for i to 100 do
>       for j to 11 do
>           t[(i-1)*11+j] := 1./(i+y[j]);
>       end do;
>   end do;
>   gt := map( G, t ):
```

Let us look at the last few entries in that list, to see whether the y-range spans the entire graph.

```
>   gt_0 := map( G, t[1100-7..1100] );
```

$$gt\_0 := [.3000000, .4000000, .5000000, .6000000,$$
$$7000000, .8000000, .9000000, .9999999]$$

That shows that the map appeared to work; we now will get complete coverage of each interval. It also showed how to refer to several elements of a Vector or a Matrix at once.

The following shows how to use `zip` to convert two separate lists of data points into a format acceptable for Maple's plotting facilities.

```
>  pts := zip( (x,y)->[x,y], t, gt ):
```

```
>  whattype( pts );
```

$$Vector_{row}$$

```
>  plot( convert(pts,list), view=[0..1,0..1],
>         colour=black );
```

This plot is shown in Figure 2.26. It is better than before, but still not perfect. There are some gaps there (probably rounding errors), and lines jumping across the discontinuities. I want no extraneous lines. For some plots with discontinuities you can use the option `discont=true` to tell Maple that the function is discontinuous and not to connect the graph across the discontinuities. However, this plot is too complicated for that approach to work. Instead, one way to fix this here is to create a whole bunch of plots, on the same graph.

```
>  pieces := array(1..100):
```

```
>  for i to 100 do
>      pieces[i] := NULL;
>      for j to 11 do
>          pieces[i] := pieces[i], 1./(1+y[j]), y[j];
>      end do;
>      pieces[i] := [pieces[i]];
>  end do:
```

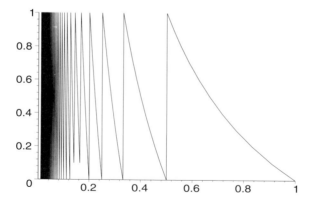

Figure 2.26: An improved plot of $G(t)$

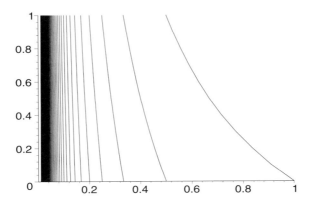

Figure 2.27: The best plot of $G(t)$

That program uses a loop to create an expression sequence, and then converts that expression sequence to a list. That construct is *inefficient*, because adding an entry to an expression sequence requires searching the whole expression sequence. Thus the above loop takes $1 + 2 + 3 + \cdots + n = n(n + 1)/2 = O(n^2)$ operations to create a list of length n. It is not crucial in this example because there are only 100 pieces, each of length 11, but it would have been better to use seq in the inner loop:

```
>   pieces := array(1..100):
>   ys := [seq(y[j],j=1..11)]:
>   for i to 100 do
>       pieces[i] := zip( (i,j)->[i,j],
>           [seq(1./(i+y[j]),j=1..11)], ys );
>   end do:
```

This code is faster than the previous by a factor of about 5. Now, to plot the pieces, we put them in a set, as follows.

```
>   plot( {seq(pieces[k],k=1..100)}, view=[0..1,0..1],
>       colour=black );
```

This plot is shown in Figure 2.27. Now *this* plot looks acceptable. There is still a little blank space next to the y-axis, because we have plotted the pieces only up to $n = 100$, and this leaves 1% of the graph uncovered. We will see yet another plot of this function, on a torus, in a subsequent section.

Remark. After having done all that, it is clear that we could have done a similar job of plotting with much less effort, by simply reparameterizing the curve into pieces: $(x, y) = (1/(n + t), t)$ as t runs from 0 to 1. Thus the command

```
>   plot( {seq( [1/(n+t), t, t=0..1], n=1..100)},
>      x=0..1, y=0..1, colour=black );
```

ought to produce nearly the same graph as before. We will not do this plot now, but rather reserve this idea for use with the torus plot later.

Polar coordinate plots

In [5], Michael W. Chamberlain draws the polar graphs of $r = \cos 5\theta + n \cos \theta$, for $0 \leq \theta \leq \pi$, for integers $n = -5$ (which gives a heart shape) to $n = 5$ (which gives a bell shape). We try to reproduce this in Maple.

```
>   restart;
>   with( plots, polarplot );
```
$$[polarplot]$$
```
>   polarplot( {seq(cos(5*theta)+n*cos(theta),n=-5..5)},
>         theta=0..2*Pi, colour=black, axes=BOXED,
>         scaling=CONSTRAINED );
```

That plot, shown in Figure 2.28, looks like the picture in [5]. We now do this another way, using a parametric plot.

```
>   r := (n,theta) -> cos(5*theta) + n*cos(theta);
```
$$r := (n, \theta) \rightarrow \cos(5\theta) + n\cos(\theta)$$

We can *animate* this with the following command, though in order to show an animation in this book I would have to resort to printing each frame on a page and asking you to flip the pages.

```
>   plots[display](
>      [seq(plot( [r(n,t)*cos(t),r(n,t)*sin(t),t=0..2*Pi]),
>         n=-5..5)], insequence=true );
```

That plot is shown in Figure 2.29.

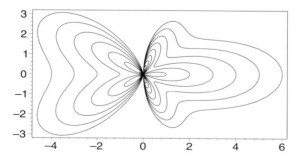

Figure 2.28: "Heart to Bell"

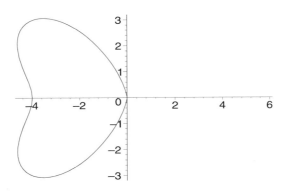

Figure 2.29: The start of an animation of "Heart to Bell," plotted parametrically

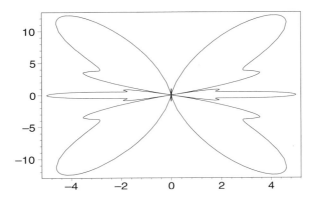

Figure 2.30: $r = (4\cos 3\theta + \cos 13\theta)/\cos\theta$

We now explore some other polar plots from an article by Temple H. Fay [20].

```
>   polarplot( (4*cos(3*theta)+cos(13*theta))/cos(theta),
>           theta=0..2*Pi, colour=black, axes=BOXED );
```

See Figure 2.30.

```
>   polarplot( (4*cos(theta)+cos(9*theta))/cos(theta),
>           theta=0..2*Pi, colour=black, axes=BOXED );
```

See Figure 2.31. Now try the *Fay butterfly*:

```
>   polarplot(exp(cos(theta))-2*cos(4*theta)+sin(theta/12)^5,
>           theta=0..24*Pi, colour=black, axes=BOXED );
```

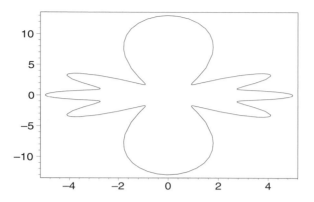

Figure 2.31: $r = (4\cos\theta + \cos 9\theta)/\cos\theta$

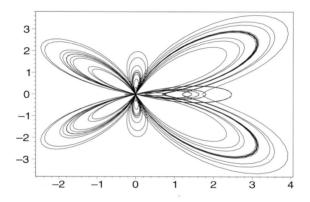

Figure 2.32: The Fay butterfly $r = \exp(\cos\theta) - 2\cos 4\theta + \sin^5(\theta/12)$

See Figure 2.32. One can select the `axes=BOXED` option from the menu of the plot, rather than redisplaying the curve with a Maple command, if desired.

Exercises

1. Plot $1/\Gamma(x)$ on $-4 \le x \le 2$.

2. What will `plot(sin, -Pi/2..Pi/2)` produce?

3. Plot $\sin x$, $\cos x$, and $\tan x$ on the same graph, on $-4\pi \le x \le 4\pi$. Use `view` to get a good scale.

4. Plot the folium of Descartes, which is given parametrically by

$$x = 3at/(1+t^3), \qquad y = 3at^2/(1+t^3).$$

Nondimensionalize first, of course, and plot x/a versus y/a.

5. Choose four random points in the x-y plane, fit a straight line to them (in the least squares sense), and plot the points and the line on the same graph. See ?regress or leastsqrs.

6. Plot $r = (4\cos m\theta + \cos n\theta)/\cos\theta$ for some odd values of m and n. If one of the values of m or n is even, the plot is supposed to look quite different. See [20] and the reference therein for more details.

7. Plot the cissoid of Diocles, whose rectangular equation is

$$y^2 = \frac{x^3}{2a - x}$$

and whose parametric equations are $x = 2a\sin^2\theta$, $y = 2a\sin^3\theta/\cos\theta$.

8. Plot representatives of the ovals of Cassini. The polar equation is $r^4 + a^4 - 2a^2r^2\cos 2\theta = b^4$. There are three qualitatively different curves, depending on whether b/a is less than 1, equal to 1, or greater than 1.

2.7.2   Three-Dimensional Plots

We continue with a plot of the Gauss map on a torus. The basic idea of this is that we want to take our flat graph, Figure 2.27, wrap it up in a tube, and then bend that tube around into a torus shape. Analytically, we are considering G as a map from the unit circle S^1 to itself; $G : S^1 \to S^1$ by $G(\exp(2\pi it)) = \exp(2\pi i/t)$, and now the "fractional part" of the map is taken care of automatically. If we use the parameterization idea from the flat graph, what we get is the following.

First we define a torus, by specifying its centreline:

```
>   restart;
>   sp := [rho*cos(2*Pi*t), rho*sin(2*Pi*t), 0, radius=b]:
```

Now let us define each piece of the curve (one "wrap," if you like), by

```
>   pc := n -> [ (rho-r*cos(2*Pi*t))*cos(2*Pi/(n+t)),
>             (rho-r*cos(2*Pi*t))*sin(2*Pi/(n+t)),
>             -r*sin(2*Pi*t)]:
```

The sign of the last (z) component was chosen to make the graph agree with the cover of the March 1992 *American Mathematical Monthly* [7]. Now we need to set suitable values for the parameters; in particular, we need to make the radius of the Gauss map slightly larger than the radius of the torus, so the hidden-line removal algorithms don't destroy it.

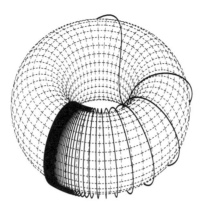

Figure 2.33: The Gauss map, graphed on a torus

```
>   (rho, r, b) := 2, 1.1, 1;
```

$$\rho, r, b := 2, 1.1, 1$$

Now we do the plots by using routines from the `plots` package. Do not worry about the warning that arises when we issue the command `with(plots)`. It is just stating that the routine for changing coordinate systems in algebraic expressions is no longer available, being shadowed by a routine of the same name for changing coordinate systems in plots.

```
>   with(plots):
```

```
Warning, the name changecoords has been redefined
```

The following generates the thickened curves, but does not display them. If we used a semicolon (;) we would see the plot structure line-printed, which is not what we want. The view is chosen (after some experimentation) to give a good scaling for the torus.

```
>   vw := [-3.1..3.1,-3.1..3.1, -2.2..2.2];
```

$$vw := [-3.1..3.1, -3.1..3.1, -2.2..2.2]$$

```
>   s := spacecurve( {seq(pc(k),  k=1..50)}, t=0..1,
>           thickness=3, colour=black, view=vw ):
```

This generates the torus defined with the centreline sp.

```
>   s2 := tubeplot( sp, t=0..1, tubepoints=30, colour=black,
>           view=vw, style=HIDDEN,
>           linestyle=3, thickness=2 ):

>   display( {s,s2}, scaling=CONSTRAINED,
>           orientation=[10,50] );
```

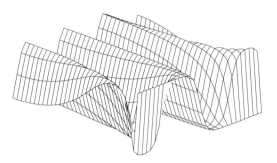

Figure 2.34: The Jacobian elliptic function sn(x, y)

After the plot has been displayed, one can use the menus to fiddle with the lighting schemes, etc., to produce the final version of the plot, instead of choosing these on the command line.

For another example, let us consider graphing sn(u, k), which is one of the Jacobian elliptic functions. We plot sn(u, k) on a compact x-interval because the convergence of sn(u, k) to tanh(x) as $k \to 1^-$ is not uniform.

```
>   restart;
>   plot3d( JacobiSN(x,y),x=-10..10, y=0..0.999999,
>           grid=[30,30], colour=black, style=HIDDEN );
```

This graph is shown in Figure 2.34. Jon Borwein remarked that this is a "graphical proof of the fast computability of the elementary functions," because it is known that the elliptic functions are quickly computable, and by continuation so is tanh(u).

The fact that sn(u, k) goes to tanh(u) is not really so evident in Figure 2.34; all the details and excitement of the limit happen for k close to 1. We can expand the interesting portion of that graph by plotting logarithmically in k, as follows.

```
>   plot3d( JacobiSN(x,1-10^(-y)),x=-10..10, y=0..6,
>           grid=[30,30], colour=black, style=HIDDEN );
```

Figure 2.35 shows the limiting case more clearly than the original figure.

Riemann surfaces

Cleve Moler noticed some time ago that Problem Solving Environments (PSEs) such as MATLAB (or Maple) could be used to plot Riemann surfaces. This provides a much-needed facility for people to learn what such objects look like. Here is a single example, the Riemann surface for arcsin z. The plot is much better seen live in Maple where the colours, and more importantly the ability of the user to

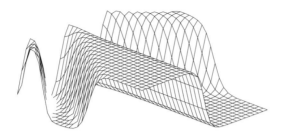

Figure 2.35: The Jacobian elliptic function $\text{sn}(x, 1 - 10^{-y})$

rotate the plot by dragging it with the mouse, allow the user to gain a real grasp of what the surface is like. See [16] for more examples, and for details of what has to be proved in each case before we can accept the picture as being realistic.

```
>   w := u + I*v;
```
$$w := u + I\,v$$
```
>   z := sin( w );
```
$$z := \sin(u + I\,v)$$
```
>   x := evalc( Re(z) );
```
$$x := \sin(u)\cosh(v)$$
```
>   y := evalc( Im(z) );
```
$$y := \cos(u)\sinh(v)$$
```
>   plot3d( [x,y,u], u=-6..6, v=-6..6, grid=[50,50],
>           colour=v, style=PATCHNOGRID, axes=NONE,
>           scaling=CONSTRAINED, orientation=[30,84],
>           view=[-6..6,-6..6,-6..6] );
```
See Figure 2.36.

Exercises

1. Use the plot3d command to plot $w = \Re(z^n)$, where $z = x + iy$ and $n = 3, 4, 5,$ and 6. Rotate the plots and see whether you can understand the pattern. Do the same for the imaginary parts of z^n.

2. Plot all the examples from ?plot3d.

3. Plot the following functions.

   (a) $|x| + |y|$
   (b) $y^2/4 - x^2/9$

Figure 2.36: A portion of the Riemann surface for arcsin x

(c) $\left(y^2 - x^2\right) / \left(y^2 + x^2\right)$

(d) $\sin\left(\sqrt{1 - x^2 - y^2}\right) / \sqrt{1 - x^2 - y^2}$

(e) $\left(x^3 y - x y^3\right) / \left(x^2 + y^2\right)$

(f) $e^{-y} \cos x$

(g) $y^2 - y^4 - x^2$

(h) $1 / \left(x^2 + 4y^2\right)$

(i) $x y^2 / \left(x^2 + y^2\right)$

(j) $1 + \cos\left(x^2 + y^2\right)$

(k) $2xy / \left(x^2 + y^2\right)$

(l) $\left| |x| - |y| \right| - |x| - |y|$

(m) $x = r \cos\theta$, $y = r \sin\theta$, $z = \cos(m\theta)/J_m(\lambda r)$, where $\lambda = 11.61984117$ and $J_m(z)$ is the Bessel J function of order m. Plot this for $m = 2$ and $m = 4$ on $0 \leq r \leq 0.9$ and $0 \leq \theta \leq 2\pi$. This plot was suggested by George Labahn.

2.7.3   Contour Plots and Other Plots

We begin with some contour plots from [53]. To produce contour plots in Maple, we first load the plots package.

```
>   with( plots ):
```

Warning, the name changecoords has been redefined

```
>   f1 := sin(y-x^2-1) + cos(2*y^2-x);
```

$$f1 := \sin(y - x^2 - 1) + \cos(2 y^2 - x)$$

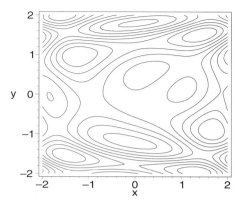

Figure 2.37: Contours of $\sin(y - x^2 - 1) + \cos(2y^2 - x)$

```
>   contourplot( f1, x=-2..2, y=-2..2, grid=[100,100],
>           colour=black, scaling=CONSTRAINED, axes=BOXED);
```

See Figure 2.37. The plot shows several local maxima, minima, and saddle points. We will give only one more contour plot here directly, but will suggest several other interesting plots in the exercises. Consider now

```
>   f2 := y + sin(x^2*y-1/x);
```

$$f2 := y + \sin(x^2 y - \frac{1}{x})$$

Actually, the paper [53] considers $f = \sin\left(y + \sin\left(x^2 y - 1/x\right)\right)$. Initially I thought that the contours of the above function should be the same as that in the paper, because if $\sin(f) = $ constant, then $f = $ constant also; hence the shape of each contour will be the same. However, the appearance of the two contour plots is quite different. Why this is so is left to the exercises.

```
>   contourplot( f2, x=-Pi..Pi, y=-Pi..Pi,
>           grid=[100,100], colour=black,
>           axes=BOXED, scaling=CONSTRAINED );
```

This plot is shown in Figure 2.38.

Exercises

1. Read the documentation for the `plots` package. There are many useful routines there that I have not covered in this book. For example, I could have used `plots[setoptions]` to shorten the example plots (by setting a common set of default options). I chose not to, for ease of reproducibility.

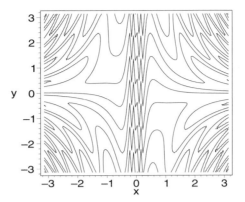

Figure 2.38: Contours of $y + \sin(x^2 y - 1/x)$

2. Plot the contours of $f = \sin(y + \sin(x^2 y - 1/x))$ and discuss why it looks different from Figure 2.38.

3. Plot the contours of

   (a) $w = \Re(z^n)$, where $z = x + iy$ and $n = 3, 4, 5,$ and 6.

   (b) $1/(x^2 + y^2 - \pi) + \exp(x + y/\pi)$ on some suitable range around the origin.

   (c) $x^4 + y^4 - 6x^2 y^2$

   (d) $(x - y)/(x + y)$

   (e) $\tan(x^2 + y)$

   (f) $\sin(x^2 - y^2 - 1) + \cos(4x^2 y^2)$

   (g) $xy + 2x - \ln(x^2 y)$

   (h) $x = r\cos\theta$, $y = r\sin\theta$, and $z = \cos(m\theta)/J_m(\lambda r)$, where $\lambda = 11.61984117$, $J_m(z)$ is the Bessel J-function of order m. Choose $m = 4$ and $m = 2$. This plot was suggested by George Labahn.

Implicit Plots

There are now three facilities for plotting implicit functions in Maple: `plots[ implicitplot ]`, `plots[ implicitplot3d ]`, and a new, experimental one, `algcurves[ plot_real_curve ]`. This new routine first uses `solve` to identify interesting points (singular points, critical points, points at minimum and maximum distance from the origin, and so on). Then it computes a numerical solu-

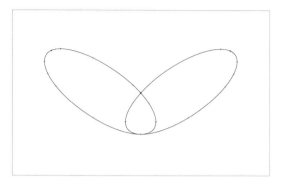

Figure 2.39: The tacnode curve as produced by `plot_real_curve`

tion of a differential equation following paths from halfway between each critical point. The numerical solution is quite fast, and when `solve` succeeds the location of critical points analytically gives a good picture. However, sometimes `solve` fails or is slow, and this means that `algcurves[ plot_real_curve ]` will fail or be slow.

```
>   restart;
```

We choose as our example the *tacnode* curve, which we take from [43]. This curve is of degree four, and contains features that make it hard (but not uncommonly hard) to plot.

```
>   tacnode := 2*x^4 - 3*x^2*y + y^4 - 2*y^3 + y^2;
```
$$tacnode := 2\,x^4 - 3\,x^2\,y + y^4 - 2\,y^3 + y^2$$

After some experimentation, we choose the colour black for the curve (so it reproduces well for this book),[9] no axes, and a particular viewing region. It should be said that the principal advantage that `algcurves[ plot_real_curve ]` has over `plots[implicitplot]` is that it chooses the region of interest automatically; by choosing the viewing region to be the same, we vitiate that advantage.

```
>   algcurves[plot_real_curve]( tacnode, x, y,
>               colorOfCurve=COLOR(RGB,0,0,0),
>               axes=NONE, view=[-2..2,-1..3] );
```

That plot is in Figure 2.39.

---

[9]It is unfortunate that the colour options for this experimental routine are not in line with other plotting functions. However, I am sure that the authors of this code will improve things for the next version of Maple. I can say that because I am one of those authors.

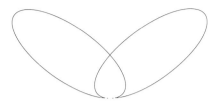

Figure 2.40: The tacnode curve by `implicitplot` (400×400)

```
>   plots[implicitplot]( tacnode, x=-2..2, y=-1..3,
>           axes=NONE, colour=BLACK, view=[-2..2,-1..3] );
```

That unsatisfactory plot is not shown.

```
>   plots[implicitplot]( tacnode, x=-2..2, y=-1..3,
>           grid=[100,100], axes=NONE,
>           colour=BLACK, view=[-2..2,-1..3] );
```

That (less unsatisfactory) plot is not shown either.

```
>   plots[implicitplot]( tacnode, x=-2..2, y=-1..3,
>           grid=[400,400], axes=NONE,
>           colour=BLACK, view=[-2..2,-1..3] );
```

That (still unsatisfactory) plot is shown in Figure 2.40. At a grid of 300 by 300, the crossing is correct. Even at 500 by 500, the origin is still not correct.

There are examples where `plots[implicitplot]` is better than `algcurves[ plot_real_curve ]`, however. See the exercises. The main advantage of `implicitplot` is that it is not limited to polynomial curves. However, as the following command shows, some portions of some transcendental curves can also be plotted by `algcurves[ plot_real_curve ]`.

```
>   algcurves[plot_real_curve]( y*exp(y)-x, x, y,
>           force=true );
```

That plot, of the Lambert W function, is in Figure 2.41.

Exercises

1. Plot the graph of the surface given implicitly by $x^3 - y^2 + z^3 = 1$. (See `?implicitplot3d`.)

Plotting the solutions of ODE

The routine `plots[odeplot]`, together with the Maple 7 implementation of good numerical methods for the solution of initial value problems for ordinary differential equations, allows reasonable plotting of such solutions.

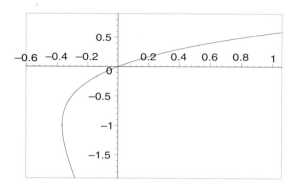

Figure 2.41: The Lambert W function by `plot_real_curve (force=true)`

The following shows how to plot a phase diagram for the Van der Pol equation

$$\ddot{x} - \epsilon(1 - x^2)\dot{x} + x = 0\,,$$

using the usual first-order system form that arises on putting $y = \dot{x}$. We choose $\epsilon = 1$ and 500. Other phase portraits are left for the exercises.

```
>   with(plots):
```

```
Warning, the name changecoords has been redefined
>   van_der_Pol := diff(x(t),t,t)
>             - epsilon*(1-x(t)^2)*diff(x(t),t) + x(t);
```

$$van\_der\_Pol := \frac{\partial^2}{\partial t^2} x(t) - \varepsilon \left(1 - x(t)^2\right) \frac{\partial}{\partial t} x(t) + x(t)$$

```
>   eq1 := eval(van_der_Pol,epsilon=1):
>   eq500 := eval(van_der_Pol,epsilon=500):
>   sol1 := dsolve( {eq1,x(0)=2,D(x)(0)=0}, x(t),
>             numeric, range=0..10, stiff=true );
```

$$sol1 := \textbf{proc}(rosenbrock\_x) \ldots \textbf{end proc}$$

```
>   plots[odeplot]( sol1, [x(t),diff(x(t),t)],
>             colour=black );
```

See Figure 2.42.

```
>   sol500 := dsolve( {eq500,x(0)=2,D(x)(0)=0}, x(t),
>             numeric, range=0..1000, stiff=true );
```

$$sol500 := \textbf{proc}(rosenbrock\_x) \ldots \textbf{end proc}$$

```
>   plots[odeplot]( sol500, [x(t),diff(x(t),t)],
>             colour=black );
```

See Figure 2.43.

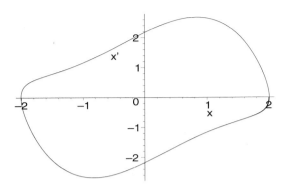

Figure 2.42: Phase portrait for the Van der Pol equation, $\epsilon = 1$

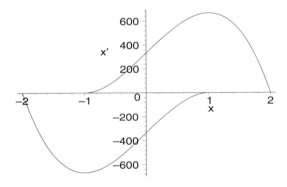

Figure 2.43: Phase portrait for the Van der Pol equation, $\epsilon = 500$

We now look at a plot of the solution to a first order differential equation, $y' = \cos(\pi t y)$, for various initial conditions. For an explanation of the curious "bunching" of the curves, see the exercises in [2]. We construct a sequence of solutions for a number of different initial conditions, as follows.

```
>   BenderOrszag := diff(y(t),t) = cos(Pi*t*y(t));
```

$$BenderOrszag := \tfrac{\partial}{\partial t}\, y(t) = \cos(\pi\, t\, y(t))$$

```
>   sols := {seq(dsolve({BenderOrszag,y(0)=k/5},
>           y(t), numeric, range=0..5 ), k=0..20 ) }:
```

We now construct a plot for each solution, and display them all together on the same graph.

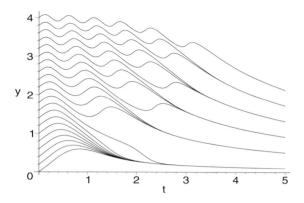

Figure 2.44: Solution of $y' = \cos(\pi t y)$ for various initial conditions

```
>   plts :={seq(
>             odeplot(sols[i],[t,y(t)],colour=black),
>             i=1..nops(sols))}:

>   display( plts );
```

See Figure 2.44.

Now let us examine a second-order differential equation system, a predator–prey model.

```
>   restart;
>   with(plots):
```

```
Warning, the name changecoords has been redefined
>   PredatorPrey := {diff(x(t),t) = -3*x(t)+4*x(t)^2
>             -x(t)*y(t)/2-x(t)^3,
>             diff(y(t),t) = -2.1*y(t)+x(t)*y(t)};
```

$$PredatorPrey := \left\{ \frac{\partial}{\partial t} y(t) = -2.1\, y(t) + x(t)\, y(t), \right.$$

$$\left. \frac{\partial}{\partial t} x(t) = -3\, x(t) + 4\, x(t)^2 - \frac{1}{2}\, x(t)\, y(t) - x(t)^3 \right\}$$

```
>   ics := [seq({x(0)=4,y(0)=k/8},k=0..32)]:
>   sols := {seq( dsolve( PredatorPrey union ics[i],
>             {x(t),y(t)}, numeric,
>             range=0..100 ),
>             i=1..nops(ics)) }:
```

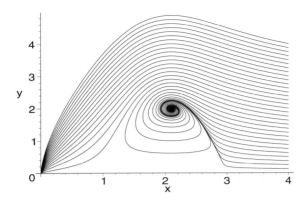

Figure 2.45: Phase plane solutions to predator–prey equations

```
> plts := {seq(
>           odeplot(sols[i],[x(t),y(t)],colour=black),
>           i=1..nops(sols)) }:

> display( plts );
```

The plot shown in Figure 2.45 agrees with other reliable numerical solutions of this equation, and with the linear stability analysis of the fixed points.

Exercises

1. Do a linear stability analysis of the predator–prey equations above near the unstable equilibrium $x = 1$, $y = 0$. You should find that the unstable manifold has the equation $x = 1 + 5u/31$, $y = u$ for small u. Use $u = 0.1$ to get a point near to the unstable equilibrium, and integrate the predator–prey equations on the range $-5 \le t \le 0$, i.e., backwards, (specify `range = -5..0` and `start=0` in the call to `dsolve`) to get a graph of the curve separating flows that tend to the nonzero population equilibrium from the ones that tend to zero. Plot the unstable manifold and the separatrix together with the plots of solutions above.

2. Plot a dodecahedron.

3. Plot $y = \int_0^x 1/\Gamma(t)\,dt$ on $-4 \le x \le 2$, first by direct use of plot and then by plotting the solution of the differential equation $y' = 1/\Gamma(x)$ using `plots[odeplot]`. Which is faster?

4. Plot the curve defined by $x^5 + y^5 = 5ax^2y^2$, first by using `implicit-plot` or `plot_real_curve` (nondimensionalize, of course, so you don't have to deal with the a), and then by putting $y = tx$ and deriving a pair of parametric equations, $x = f(t)$ and $y = tf(t)$, and plotting those. See Section 3.9.1.

5. The continuous logistic equation is $\dot{x} = x(1-x)$, $x(0) = \alpha$. Solve this analytically in Maple and plot the solution for $\alpha = 0, \frac{1}{5}, \frac{2}{5}, \ldots, \frac{9}{5}, 2$. Compare the time it takes to do this with the time taken to generate the plot by use of `dsolve[numeric]` and `plots[odeplot]`.

6. The discrete logistic equation is $x_{n+1} = \mu x_n(1 - x_n)$, $x_0 = \alpha$. It arises, for example, by applying Euler's method to the continuous logistic equation, but also directly in applications. Write a Maple program to approximate the attractive periodic points of this map, given the numerical value of μ. For values of μ varying between 0 and 4, plot these periodic points.

7. Learn how to use `textplot` and `textplot3d` to annotate your plots.

8. Use `conformal` to examine the complex transformations $w = \ln(z)$ and $w = (z^2 - 1)^{1/2}$.

9. Plot a pentagon and a five-pointed star on the same graph. See `polygon-plot`, and use $[\cos(2\pi k/5), \sin(2\pi k/5)]$ to generate the coordinates of the vertices of the pentagon and a similar sequence for the five-pointed star.

2.7.4 Common Errors

Probably the single most common error is *believing what you see*. Computer-generated plots can be misleading, even if the software is doing what it is supposed to. Sometimes the misleading behaviour is obvious (plot, for example, $\sin(x^2)$ on $0 \le x \le 30$, or $\sin(1/x)$ on any interval containing 0). At other times, it is not obvious. Plot

$$x^7 - \frac{7x^6}{2} + \frac{161x^5}{32} - \frac{245x^4}{64} + \frac{6769x^3}{4096} - \frac{3283x^2}{8192} + \frac{3267x}{65536} - \frac{315}{131072}$$

on $-1 \le x \le 2$ and note that it looks completely flat on $0 \le x \le 1$. However, there are seven real roots equally spaced on this interval, and thus six extrema. This fact is not at all evident from the graph, on this scale. In fact, automatic scaling of polynomials so that interesting features can be seen is a difficult problem, and I have not yet seen a program that can do it satisfactorily.

Common Maple mistakes include:

1. Forgetting that x has a value.

```
>   restart;
>   x := 17;
```
$$x := 17$$

Now a long session, in which we forget that x has a value. Then, finally, we issue the command

```
>   plot( sin(x), x=0..5 );
```

```
Error, (in plot) invalid arguments
```

To fix this problem, unassign x by x := 'x'; and reissue the plot command.

2. Plotting a function defined by a procedure that expects only numerical arguments without preventing premature evaluation. Suppose for example we wished to plot a piecewise-defined function defined by the following.

```
>   p := proc( t );
>       if t > 1 then
>           sin(Pi*t) + 1
>       else
>           t^2
>       end if
>   end proc:
>   plot( p(x), x=0..2 );
```

```
Error, (in p) cannot evaluate boolean: -x < -1
```

This error message arises because p has been evaluated at the symbolic argument x, and Maple can't tell whether $x > 1$ or not. One way to fix it is to use quotation marks to prevent premature evaluation.

```
>   plot( 'p(x)', x=0..2 );
```

Another way is to plot using the operator syntax. This technique is highly recommended.

```
>   plot( p, 0..2 );
```

Still another way is to rewrite the procedure to return unevaluated if the argument is not numeric.

```
>   p := proc(t);
>       if not type(t, numeric) then
>           'p'(t)
>       elif t > 1 then
```

```
>            sin(Pi*t) + 1
>      else
>           t^2
>      end if
>   end proc:

>   plot( p(x), x=0..2 );
```

3. Finally, trying to plot functions that contain symbols is impossible. For example, one cannot plot $r = a \cos \theta$ in polar coordinates. One should nondimensionalize and plot r/a versus θ instead.

Exercises

1. Plot a logarithmic spiral $r = \exp(\theta)$. Does the plot show the qualitative features correctly?

2. Plot $(\sin^2 x + \cos^2 y - 1)/(\tan^2 x + 1 - \sec^2 y)$ in three dimensions. What happens as $y \to x$?

3. (From Bill Gosper) Plot $x \ln |x|$. Does it look vertical at the origin? Should it? Try a sequence of plots on the range $[-10^{-k}, 10^{-k}]$ for various integers k. Try it again with the option `scaling=CONSTRAINED`.

2.7.5   Getting Hard Copy of your Plots

One way to obtain hard copy of a plot is simply to print your worksheet. Printing from a windowing session is usually as easy as finding the correct menu item: Export under the File menu puts a worksheet into different formats, including LaTeX and HTML; and that is how most of the graphs were generated for this book.

However, you may wish to get a file copy of just the plot. The easiest way to do this in Maple 7 is to right-click on the plot, and choose "Export to," which gives you the choice of several formats, including encapsulated PostScript or JPEG.

To get a hard copy in a command-line style, see `?interface[plotoutput]` or `?plotsetup` for details.

Here we show how to produce a PostScript plot directly.

The following function $B(v)$ arises in the analysis of the effect of solving $y' = y^2$ by Euler's method [8, 3]. We note that $B(v)$ satisfies the following functional equation:

$$B(v) = \frac{(1+v)^2}{1+2v} B(v + v^2) \,,$$

and has a series expansion beginning

$$B(v) = 1 - v + \frac{3}{2}v^2 + O(v^3) \,.$$

We have already seen the Maple code for generating an arbitrary number of the coefficients for this series, in Section 2.4. It is reproduced here for convenience.

```
> c := proc( n ) option remember;
>    -1/(n-1)*add( binomial(n-i+1, i+1)*c(n-i), i=1..n-1 )
> end proc:

> c(1) := 1;
```

$$c(1) := 1$$

The following uses Levin's u transform to accelerate the sum, which turns the asymptotic series into a convergent one. This works, by experiment, only if $-0.1 \le v \le 0.1$.

```
> Bseries := v->evalf(Sum('c(n)*v^(n-1)',n=1..infinity));
```

$$Bseries := v \rightarrow evalf\left(\sum_{n=1}^{\infty}{}' c(n)\, v^{(n-1)},\right)$$

We use two infinite product representations to compute B for large values of v.

```
> restart;
> B := proc(v) local p,v0,u0;
>     if not type(v, numeric) then
>         'B(v)'
>     elif v < -1 then
>         (1.+v)^2/(1.+2*v)*B(v*(v+1.))
>     elif v=-1 then
>         0.
>     elif -1 < v and v < 0 then
>         v0 := v;
>         p  := 1.;
>         while v0 < -0.1 do
>             p  := p*(1.+v0)^2/(1+2.*v0);
>             v0 := v0*(v0+1.);
>         end do:
>         p*Bseries(v0)
>     elif v = 0 then
>         1.
>     else
>         u0 := v;
>         p  := 1.;
>         while u0 > 0.1 do
>             u0 :=  2.*u0/(1.+(1.+4.*u0)^(1/2));
>             p  := p*(1.+2.*u0)/(1+u0)^2;
>         end do:
>         p*Bseries(u0)
>     end if:
> end proc:
```

At last, the commands to actually produce the plot.

```
>   plotsetup(ps, plotoutput='Beyn.ps',
>           plotoptions='portrait, noborder');
```

```
>   plot( B, -2..2, -2..2, discont=true, colour=black );
```

This function is difficult to graph in any lower-level language.

The above session produces in the file "Beyn.ps" the PostScript commands shown in Figure 2.46. Sending the file to a PostScript printer produces the plot shown in Figure 2.47. Sometimes it is necessary to alter the bounding box (what Maple produces is actually Encapsulated PostScript) or the drawborder flag from true to false.

Remark. If you start Maple by double-clicking on a worksheet, then the "current directory" will be the folder that the worksheet was in. If you start Maple by

```
%!PS-Adobe-3.0 EPSF-2.0
%%Title: Maple plot
%%Creator: Maple
%%Pages:  1
%%BoundingBox: 110 131 496 660
%%DocumentNeededResources: font Helvetica
%%EndComments
20 dict begin
gsave
/drawborder false def
/m {moveto} def
<****** many PostScript commands omitted ******>
% The following draws a box around the plot,
% if the variable drawborder is true
drawborder {
/bd boundarythick 2 idiv def
[] 0 setdash
NP bd bd m bd 6923 bd sub l
5000 bd sub 6923 bd sub l
5000 bd sub bd l
bd bd l S
} if % end of if to draw the border

showpage
grestore
end
%%EOF
```

Figure 2.46: The PostScript commands in the output file used to print Figure 2.47

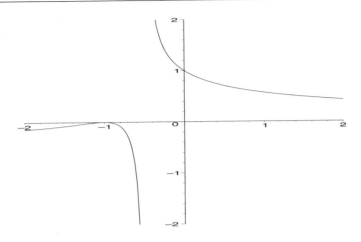

Figure 2.47: The graph of $B(v)$

clicking on Maple's desktop icon, then the "current directory" will be Maple's bin directory; the effect of this on the following command is that the plot will be put in the bin directory. In a well-set-up installation using a real operating system you will not have write access there. So, use the full path name in the `plotoutput` option if you aren't sure where the plot will go, or else use the `currentdir` command to set the path.

2.8   Packages in Maple

There were fifty-three packages of Maple library functions at the time of this writing. See ?packages for a list. By Pareto's principle, one can expect to use only 20% of these packages for 80% of your work; this translates to about 10 of them, and this reflects my own experience. Of course, the 10 most useful to you may well be different from the 10 most useful to me. For me, the most useful packages are

1. `LinearAlgebra`

2. `plots`

3. `Groebner`

4. `codegen`

5. `Matlab`

6. inttrans

7. student

8. Units (I am anticipating—this is a new one)

9. MathML (again, a new one)

10. numapprox.

LinearAlgebra, plots, Groebner, and codegen have been shown else-where in this book. Here I discuss the link to MATLAB, too little known, the numapprox package, and the two new packages Units and MathML.

2.8.1   The MATLAB Link

```
>   restart;
```

We gain access to MATLAB from Maple by using with. If this does not work on your system, check that you have MATLAB installed and that Maple is configured correctly. Issue the Maple command ?Matlab to learn how to configure your system.

```
>   with( Matlab );
```

> [*chol, closelink, defined, det, dimensions, eig, evalM, fft, getvar,*
> *inv, lu, ode45, openlink, qr, setvar, size, square, transpose*]

```
>   with( LinearAlgebra ):
>   Digits := trunc( evalhf(Digits) );
```

$$Digits := 14$$

We first give an example of using features at the time of writing that are superior in MATLAB, namely elementwise vector operations and the Fast Fourier Transform (FFT).

```
>   evalM( " x=linspace(0,pi,1024) " );
>   evalM( " v=sin(50*x)+0.1*sin(200*x)"
>          "+0.5*sin(500*x)+0.00001*rand(1,1024) " );
>   V := getvar( "v" );
```

$$V := \big[1024 \text{ Element Row Vector Data Type : float}[8]$$

$$\text{Storage : rectangular Order : Fortran\_order} \big]$$

The following command is wasteful, because v is already in MATLAB and so I don't have to pass it over again; but the following command shows how to compute the FFT of a vector computed in Maple.

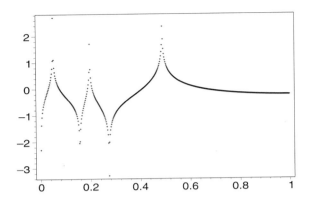

Figure 2.48: A power spectrum computed in MATLAB via the link from Maple

```
>   p := fft( V );
```

$$p := \big[\, 1024 \text{ Element Row Vector   Data Type} : \text{complex}[8]$$
$$\text{Storage} : \text{rectangular   Order} : \text{Fortran\_order}\,\big]$$

Likewise, I could compute the spectrum in MATLAB, and this means that I wouldn't have to use map, but doing it this way shows that you can manipulate in Maple objects created in MATLAB.

```
>   Spectrum := map( abs, p ):
>   plot( [seq( [2*(k-1)/1024,log[10](Spectrum[k])],
>                     k=1..512 )],
>            style=POINT, colour=black, axes=BOXED );
```

See Figure 2.48.

A more complicated example.

We here show how to use MATLAB from Maple to find an approximate minimum of a multivariate function, using MATLAB's implementation of the Nelder–Mead method. We show with this example how to build a MATLAB m-file from within Maple (of course, this is worth the effort only if you are going to use the worksheet more than once).

```
>   restart;
>   with( Matlab ):
```

The optimization problem we try to solve is to find the minimum perturbation of two polynomials that allows them to have a nontrivial GCD. This is part of an active research area now, called *Symbolic-Numeric Algorithms for Polynomials* (SNAP). Here we just use a degree 3 and degree 2 polynomial as examples.

```
>  p := expand( (t-1.99)*(t-3.0)*(t-1) );
```

$$p := t^3 - 5.99\, t^2 - 5.970 + 10.960\, t$$

```
>  q := expand( (t+1)*(t-2.01) );
```

$$q := t^2 - 1.01\, t - 2.01$$

```
>  N := 3;
```

$$N := 3$$

We use an *inert function*, namely $x(i)$, in Maple as a placeholder for what will be a MATLAB array. This is a trick.

```
>  p1 := p + add( x(i)*t^(i-1), i=1..N );
```

$$p1 := t^3 - 5.99\, t^2 - 5.970 + 10.960\, t + x(1) + x(2)\, t + x(3)\, t^2$$

```
>  q1 := q + add( x(i+N)*t^(i-1), i=1..N-1 );
```

$$q1 := t^2 - 1.01\, t - 2.01 + x(4) + x(5)\, t$$

We will require that the perturbation be such that it makes the two polynomials have a common root. One method to do this is to add a constraint and penalize the cost function if the constraint is violated.

```
>  constraint := resultant( p1, q1, t );
```

$$
\begin{aligned}
constraint :=\ & -2.\, x(3)\, x(4)\, x(1) + 1.\, x(2)\, x(5)^2\, x(4) \\
& - 1.\, x(2)\, x(1)\, x(5) + 3.97\, x(2)\, x(5)\, x(4) \\
& + 1.\, x(3)\, x(5)^2\, x(1) + 2.01\, x(3)\, x(2)\, x(5) \\
& - 6.940\, x(3)\, x(4)\, x(5) - x(5)\, x(4)^2\, x(3) + 3.\, x(4)\, x(1)\, x(5) \\
& + 1.01\, x(3)\, x(2)\, x(4) - 2.02\, x(3)\, x(1)\, x(5) \\
& - 23.939998\, x(1) - 48.059202\, x(2) - 96.6591900\, x(3) \\
& - 8.1152120\, x(4) - 15.7928280\, x(5) \\
& - 1.\, x(3)\, x(2)\, x(5)\, x(4) + 67.10900\, x(4)\, x(3) \\
& - 7.9505\, x(1)\, x(5) - 2.0097\, x(2)\, x(5) + 24.9302\, x(2)\, x(4) \\
& + .47835220 + 1.\, x(2)^2\, x(4) - 2.\, x(2)\, x(4)^2 + 1.01\, x(2)\, x(1) \\
& + 8.95\, x(4)\, x(1) + 1.52140\, x(4)\, x(5) - 2.96\, x(5)^2\, x(1) \\
& - 10.97\, x(4)^2\, x(3) - 5.970\, x(3)\, x(5)^2 + 30.04890\, x(3)\, x(5) \\
& - 4.35840\, x(5)^2 + 1.8802\, x(4)^2 - 2.01\, x(2)^2 + 4.0401\, x(3)^2 \\
& + 1.\, x(1)^2 + x(4)^3 + 5.970\, x(5)^3 + 5.0401\, x(3)\, x(1) \\
& - 4.02\, x(4)\, x(3)^2 + 1.\, x(4)^2\, x(3)^2 - 2.0301\, x(3)\, x(2) \\
& - 1.\, x(5)^3\, x(1) + 5.99\, x(5)\, x(4)^2 - 2.01\, x(2)\, x(5)^2 \\
& + 10.960\, x(4)\, x(5)^2
\end{aligned}
$$

The penalty factor here is huge (but as we will see later it doesn't work).

```
>    unconstrained := add( x(i)^2, i=1..2*N-1 )
>             + 1.0e30*constraint^2:
```

Here is how to build an *m*-file "in a bottle." We use the formatted print statement fprintf to transfer our expression containing $x(1)$, $x(2)$, and so on, to MATLAB. The fopen and fclose commands are simple. The fprintf command returns the length of the line it wrote.

```
>    fd := fopen( "C:/books/ess/programs/unconstrained.m",
>             WRITE );
>    fprintf( fd, "function y = unconstrained(x)\n" );
>    fprintf( fd, "y=%a;\n", unconstrained );
>    fprintf( fd, "end;\n" );
>    fclose( fd );
```

$$fd := 0$$
$$30$$
$$813$$
$$5$$

Remember that MATLAB expects its file names to use backslashes, but that backslash is the Maple string escape character (used above to put newlines into the file). Therefore, we must double them up in the string that we pass to evalM.

```
>    evalM( "path('C:\\books\\ess\\programs',path)" );
>    initialguess := [seq(0.,k=0..2*N-1)];
```

$$initialguess := [0., 0., 0., 0., 0., 0.]$$

We can build the command string to contain a Maple result by using sprintf, but we must be aware that the maximum length of string we can pass to MATLAB via evalM is 255 characters. This is, I believe, undocumented in Maple 7, but may well be increased in future versions.

```
>    mstring := sprintf( "pert = fminsearch( @unconstrained,"
>             " %a, optimset('TolX',1.0e-14) )",
>             initialguess ) ;
```

$$mstring := \text{"pert = fminsearch(@unconstrained,}$$
$$[0., 0., 0., 0., 0., 0.],$$
$$optimset('TolX',1.0e-14))\text{"}$$

Notice that there is no comma between the two strings input above; this uses the implicit string concatenation feature to allow us to build up a one-line string over two short lines and thus still fit on the page. I broke the lines of the output to fit, also.

```
>    evalM( mstring );
```

That command executed *very* quickly. But, in this case, it didn't happen to give us anything useful (but the point of this example is to show how to pass commands back and forth between Maple and MATLAB, and this is now evident).

```
>  ans := getvar( "pert" );
```

$ans := [-.119075316201159996\ 10^{-7},\ -.238144846575143342\ 10^{-7},$
$-.476277966373473828\ 10^{-7},\ .0000855183747847056202,$
$.0000593128369391826078,\ -.0000463488206110123874]$

```
>  eval( [unconstrained,constraint],
>          [seq(x(i)=ans[i],i=1..2*N)] );
```

$[.2272691301\ 10^{30},\ .4767275219]$

```
>  eval( [unconstrained,constraint],
>          [seq(x(i)=initialguess[i],i=1..2*N)] );
```

$[.2288208272\ 10^{30},\ .47835220]$

```
>  perturbed := eval( [p1,q1],
>          [seq(x(i)=ans[i],i=1..2*N)] );
```

$perturbed := [t^3 - 5.990000048\,t^2 - 5.970000012 + 10.95999998\,t,$
$t^2 - 1.009940687\,t - 2.009914482]$

```
>  fsolve( perturbed[1], t, complex );
```

$1.000000040,\ 1.989999758,\ 3.000000250$

```
>  fsolve( perturbed[2], t, complex );
```

$-.9999912938,\ 2.009931981$

The point of this vignette is to show how to build *m*-files, pass them to MATLAB through the `Matlab` link, automatically generate command strings, and use the results in Maple. Happy MATLAB-ing!

2.8.2 numapprox

The numapprox package allows computation of best approximations to functions. Here we suppose that we encounter the Lerch Φ function and wish to construct a formula for efficient numerical evaluation (because, say, we need to evaluate it millions of times in a numerical language such as C).

```
>  with( numapprox ):
```

We get help on the `minimax` function, which computes a best approximation using the Remez algorithm, by issuing the following command.

```
>  ?minimax
```

This section defines our sample problem:

```
>   assume( n > 0 );
>   F := int( 1/(1+x^n), x=0..t );
```

$$F := \frac{t\,\text{LerchPhi}\left(-t^n,\,1,\,\dfrac{1}{n}\right)}{n}$$

with assumptions on n

```
>   f3 := eval( F, n=3 );
```

$$f3 := \frac{1}{3}\,t\,\text{LerchPhi}\left(-t^3,\,1,\,\frac{1}{3}\right)$$

We work to hardware float precision, although it might be a good idea to work to higher precision here in order to construct an approximation valid to full machine accuracy. In this example we assume that an approximation valid to single precision is "good enough".

```
>   Digits := trunc( evalhf(Digits) );
```

$$Digits := 14$$

This command computes the approximation.

```
>   f3approx := minimax( f3, 0..1, [5,5], 1, 'maxerror' );
```

$$\begin{aligned}
f3approx := x \rightarrow\ & (.280437504\,10^{-6} + (1.3457701073709 + (\\
& -1.2630289578807 \\
& + (.95862028992109 - .38663694199434\,x)\,x)x)x)/(\\
& 1.3458133661670 + (-1.2641467040383 + (\\
& .96980628529424 \\
& + (-.10602438251847 - .16195566297938\,x)\,x)x)x)
\end{aligned}$$

```
>   maxerror;
```

$$.20837906\,10^{-6}$$

```
>   plot( f3-f3approx(t), t=0..1 );
```

See Figure 2.49. We see the typical equioscillation property of best approximation errors.

2.8.3 Units

My first actual use of the Units package was in the preparation of this book. My production editor told me that the width of the text should be 27 picas, and I obediently used that. However, there is an intermittent bug in the LaTeX macros supplied with Maple 7 for handling Export to LaTeX from a Maple worksheet, and *sometimes* the line widths in this book were not right. Without measurement, I couldn't tell whether some lines were too long or others were too short. So I asked Maple to tell me what length 27 picas was in centimetres, so that I could check with a ruler.

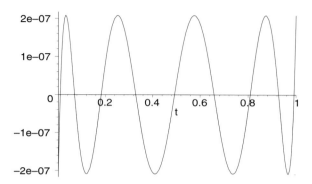

Figure 2.49: The error in approximating the Lerch Φ function by a minimax $[5, 5]$ approximant on $0 \le x \le 1$

```
> restart;
> Digits := 5;
```
$$Digits := 5$$
```
> width := convert( 27.0, units, pica, cm );
```
$$width := 11.387$$
```
> convert( 12.0, units, inch, pica );
```
$$72.270$$

This enabled me to find what was wrong (some lines were too short), and to find a workaround. I conclude that the `Units` package will be very useful. I give only some playful further examples below.

```
> restart;
> convert(0., temperature, Celsius, kelvin);
```
$$273.1500000$$

It's probably a bug that you have to capitalize `Celsius` but you are not allowed to capitalize `kelvin` (Maple is in accordance with the relevant standards: my contention is just that Maple should allow you to choose to capitalize or not, as a convenience). But, I have another shaggy dog to pursue here.

In my home town of Prince George, British Columbia, it has (three times in my memory) reached a temperature of $-55°$ Celsius.

```
> convert( -55., temperature, Celsius, Fahrenheit );
```
$$-67.0000000$$

Or, maybe, only $-55°$ Fahrenheit.

```
>  convert( -55., temperature, Fahrenheit, Celsius );
```
$$-48.3333334$$

It's still cold. I have a story, from John D. Corless (my father), about the last time it did this. It seems that birds (that weren't smart enough to have migrated before it got cold) cluster around chimneys on houses when it gets that cold. If they are overcome by smoke, they fall off the chimneys. Sometimes they fall outside, and sometimes inside. If they fall inside, and survive the fire at the bottom, they often escape the fireplace and fly around inside the house, getting soot all over. This story was not believed, at first, by Connie Corless (one of my sisters-in-law, also then a resident of Prince George). My father, twenty minutes after telling this story to her, confounded her disbelief by bringing up a blackbird from the basement (well, it was really a starling, but covered in soot from being inside the wood stove, having come down the flue). At any rate, Dad reports that the bird "looked at me reproachfully as I put it outside again."

This was the coldest week in a whole month of temperatures below $-40°$, which is well known to be just as cold in either scale:

```
>  convert( -40., temperature, Celsius, Fahrenheit );
```
$$-40.0000000$$

Then it warmed up to $-20°$ C. "It was so warm, I could push a shopping cart across the parking lot without wearing gloves." So you now know that $-20°$ C is actually *warm*. It never gets colder than this in London, Ontario, where I now live.

```
>  convert( -20., temperature, Celsius, Fahrenheit );
```
$$-4.0000000$$

In my youth, it never got warmer than about $86°$ F in Prince George.

```
>  convert(  86., temperature, Fahrenheit, Celsius );
```
$$30.0000000$$

However, in London in recent years it has reached $42°$ C, with (I am sorry to say) quite a bit of humidity.

```
>  convert(  42., temperature, Celsius, Fahrenheit );
```
$$107.6000000$$

The Units package is good for other things besides temperature: for example, many people believe that the power ratings of cars should not be in the medieval horsepower units, but rather kilowatts:

```
>  convert( 165., 'units', 'Hp', 'kW' );
```
$$123.0404788$$

Other uses:

```
>   restart;
```

```
>   with( Units[Natural] ):
```

Warning, the assigned name polar now has a global binding
Warning, these protected names have been redefined and unprotected:
*, +, -, /, <, <=, <>, =, Im, Re, ^, abs, arccos, arccosh, arccot,
arccoth, arccsc, arccsch, arcsec, arcsech, arcsin, arcsinh, arctan,
arctanh, argument, ceil, collect, combine, conjugate, convert, cos,
cosh, cot, coth, csc, csch, csgn, diff, eval, evalc, evalr, exp,
expand, factor, floor, frac, int, ln, log, log10, max, min, normal,
root, round, sec, sech, shake, signum, simplify, sin, sinh, sqrt,
surd, tan, tanh, trunc, type, verify

```
>   6*ft + 4.*inches;
```

$$1.930400000 \, [m]$$

```
>   restart;
```

The speed of sound is Mach 1, which unit uses M as its symbol:

```
>   convert( 1.0, 'units', 'M', 'km/s' );
```

$$.3314600000$$

```
>   1/%;
```

$$3.016955289$$

Therefore for every kilometer away lightning strikes, there will be about three seconds between the time that you see the flash and the time that you hear the thunder.

```
>   convert( 1.0, 'units', 'M', 'mi/s' );
```

$$.2059596954$$

```
>   1/%;
```

$$4.855318892$$

If you're still using (statute) miles, you thus know that the relevant number is about five seconds per mile.

```
>   restart;
```

```
>   with(Units):
```

```
>   GetUnit(megatonne);
```

> *tonne, context = SI, default = false, conversion = 1000000 grams$_{SI}$,*
> *prefix = SI_positive, symbol = t, symbols = {t},*
> *spelling = tonne, plural = tonnes,*
> *spellings = {tonnes, tonne}, abbreviation = none,*
> *abbreviations = {}*

That was a million tonnes of matter, not a megaton as in atomic bombs.

```
>  restart;
```

How much energy is in a kilogram of matter?

```
>  convert( 1., units, kg, joule, energy=true );
```
$$.8987551787\,10^{17}$$

```
>  restart;
>  with( Units[Natural] ):
```

I suppressed the warning that time. How much energy in a kilogram of matter, again? Let's do it directly.

```
>  mass := 1.0*kg;
```
$$mass := 1.0\,[kg]$$

```
>  c := 3.00e8 * m/s;
```
$$c := .300\,10^9\,\left[\frac{m}{s}\right]$$

```
>  energy := mass*c^2;
```
$$energy := .900000\,10^{17}\,[J]$$

Now a voltage gain.

```
>  3.*V/V(base);
```
$$3.\,\left[\frac{V}{V(base)}\right]$$

```
>  ln( % );
```
$$1.098612289\,[Np]$$

```
>  convert( %, units, dB );
```
$$9.542425097\,[dB]$$

Therefore, a gain factor 3 is a gain of about 10 decibels.

```
>  min( 3*m, 10*ft, 3.2*yd);
```
$$2.926080000\,[m]$$

Minimizing makes the units consistent:

```
>  min( 3.*m, 10.*ft, 1.0*x*yd);
```
$$\min(3., .9144000000\,x)\,[m]$$

Calculus works with units:

```
>  diff( (x^2+3*x+4)*m, x*s );
```
$$(2\,x + 3)\,\left[\frac{m}{s}\right]$$

```
>  int( %, x*s=4..7);
```
$$42\,[m]$$

```
>  plot( sin(x*degrees), x=0..180 );
```

That plot is not shown here, because it is just what you would expect.

```
> floor( 9.2*s );
```

$$9\,[s]$$

```
> sin((tau*s)*radian/s);
```

$$\sin(\tau)$$

2.8.4   MathML

This section gives some simple examples of exporting Maple expressions to MathML. MathML is the new standard for representation of mathematical objects in web browsers, so that mathematical documents can be represented, resized, searched, and indexed.

The following just uses Maple to find the roots of a quadratic equation, and to produce the MathML to display them.

```
> restart;
```

```
> with( MathML );
```

$$[\textit{Export}, \textit{ExportContent}, \textit{ExportPresentation}, \textit{Import}, \textit{ImportContent}]$$

```
> p := x^2 - 2*b*x + c;
```

$$p := x^2 - 2\,b\,x + c$$

```
> s := solve( p, x );
```

$$s := b + \sqrt{b^2 - c},\; b - \sqrt{b^2 - c}$$

```
<math xmlns='http://www.w3.org/1998/Math/MathML'>
  <mrow>
    <mi>b</mi>
    <mo>+</mo>
    <msqrt>
      <mrow>
        <mfenced>
          <msup>
            <mi>b</mi>
            <mn>2</mn>
          </msup>
        </mfenced>
        <mo>-</mo>
        <mi>c</mi>
      </mrow>
    </msqrt>
  </mrow>
</math>
```

Figure 2.50: The presentation MathML that Maple emits for one solution of $x^2 - 2bx + c = 0$

```
>  r1 := ExportPresentation( s[1] ):
>  XMLTools[Print]( r1 );
```

See Figure 2.50. The output can be saved in a file, and then web browsers like Amaya (http:// www.w3.org) or WebEQ (http:// www.dessci.com/) can render it for display in a web document.

3

Programming in Maple

> [Lady Fiorinda was] in the theoric of these matters liberally grounded through daily sage expositions and informations by Doctor Vandermast, who had these four years past been to her for instructor and tutor. To try her paces and put in practice the doctor's principles and her own most will-o'-the-wisp and unexperimental embroiderings upon them, ready means lay to hand. . .
>
> —E. R. Eddison, *The Mezentian Gate*, Book VI.

Maple is useful as a collection of "black boxes," but it is more useful still as a very high-level programming language. Since most of the tasks undertaken by Maple are "one-off" calculations (as opposed to "batch" calculations, which require many executions of the same program), it makes sense that Maple is an interpreted, rather than compiled, language. This is true even for the Maple library, because for large problems the cost will be dominated by the manipulation of large objects. Some crucial operations, though, are performed by kernel routines, which *are* compiled for efficiency.

With the external calling features, new to Maple 6 and improved for Maple 7, it is now possible to "tune" these efficiency tradeoffs more closely for any particular application.

Maple procedures can be divided loosely into two types: operators, and more general procedures. Procedures and related structures may be bundled together into a *module*. An operator is meant to imitate a mathematical operator, both in notation (insofar as this is possible in ASCII) and in action. The first section of this chapter deals with general procedures and their uses. These can do essentially anything computable. Since Maple is a high-level language, you can express these actions in many ways. The section following that looks at operators.

For more in-depth information on how to program in Maple, see [44] and the detailed examples in the directory `samples/ch06`, which can be found on Windows systems in the directory `Program Files/Maple 7`. For an extended example of revising a program for efficiency, see [50]. For examples of useful programs, see [21].

3.1 Procedures

A Maple procedure always returns a value. It is the value of the last statement executed in the procedure before returning, or else the value of an explicit `return` statement. See `?return`. This value may be NULL, which does not print anything on output. The distinction between NULL and "no value" is academic. One important consequence of a procedure returning NULL is that the environment variables %, %%, and %%% are not changed. The procedures `print` and `lprint` use this deliberately. [We will discuss environment variables in more detail later.] The procedures `solve`, `fsolve`, and `dsolve`, for example, will return NULL if they find no solution, and sometimes this takes special handling in programs. One simple way to deal with this with `solve` is to enclose the results from `solve` in set braces ({ }), converting a possible NULL value to the empty set (and incidentally removing multiplicity; use list brackets if you wish to preserve multiplicity).

Examples of simple Maple procedures follow. The first procedure accepts as argument an integer n and returns the value true or false, depending on whether n is divisible by 17. The second example illustrates the use of `for`-loops.

Pick a consistent indentation style, such as below. [My editor pointed out that the following procedure breaks the rule I mentioned in the first chapter, of not naming your routines facetiously. The first routine below should be named `DivisibleBySeventeen`, or something equally informative but shorter; however, these routines don't actually do anything *useful*, and are just intended to exhibit basic features of procedures. So I feel justified in breaking my rule; which, after all, is not supposed to be "chipped in stone."]

```
>   Fred := proc( n::integer )
>      if n mod 17 = 0 then
>         do_something_17_ish;
>         do_some_more_stuff;
>         "The input was divisible by 17"
>      else
>         do_something_not_17_ish;
>         do_some_more_stuff;
>         "The input was NOT divisible by 17"
>      end if
>   end proc:
```

The last value computed in the taken branch of the `if` statement is the value that is returned by the `if` statement and hence by the procedure.

```
>   Fred( 1 );
```

$$\text{``The input was NOT divisible by 17''}$$

```
>   Fred( 17 );
```

$$\text{``The input was divisible by 17''}$$

```
>   Fred( -17 );
```

$$\text{``The input was divisible by 17''}$$

If we pass something symbolic as input to Fred, then it fails the type-checking of the arguments:

```
>   Fred( N );
```

```
Error, invalid input: Fred expects its 1st argument, n, to be
of type integer, but received N
```

> Ginger Rogers did everything Fred Astaire ever did, only backwards and in high heels.
>
> —Anonymous

```
>   Ginger := proc( x::numeric )
>      local i, j, k, s;
>      s := 0;
>      for i from 0 to 5 while evalb( s >= 0 ) do
>         for j from -1 to 3 do
>            for k to 3 do
>                s := s + x^(i+j-k);
>            end do
>         end do
>      end do
>   end proc:
```

The routine evalb evaluates to a Boolean value, one of true, false, or FAIL.

```
>   Ginger( 3 );
```

$$\frac{572572}{81}$$

```
>   Ginger( 3. );
```

$$7068.790123$$

```
>   Ginger( 1 );
```

$$90$$

```
>   Ginger( -3. );
```

$$-10.54320988$$

```
>   Ginger(-3.001);
```

$$-10.55682503$$

There is really only one loop construct in Maple. It is a generalized `for/while` loop, which can have a logical condition as well as a counter. See `?for` or `?while` for details. Finally, the very common case "for i from 1 to n do" can be abbreviated as "for i to n do," and if the index is not needed in the loop, even as "to n do."
Some error conditions:

```
>   Ginger( x );
```

```
Error, invalid input: Ginger expects its 1st argument, x, to be
of type numeric, but received x
```

```
>   Ginger( infinity );
```

```
Error, invalid input: Ginger expects its 1st argument, x, to be
of type numeric, but received infinity
```

```
>   Ginger( NULL );
```

```
Error, (in Ginger) Ginger uses a 1st argument, x (of type
numeric), which is missing
```

```
>   Ginger( 0 );
```

```
Error, (in Ginger) numeric exception: division by zero
```

It is very useful to be able to find out which statement in the procedure caused the error, by using `tracelast`:

```
>   tracelast;
```

```
 Ginger called with arguments: 0
 #(Ginger,5): s := s+x^(i+j-k)
Error, (in Ginger) numeric exception: division by zero
 locals defined as: i = 0, j = -1, k = 1, s = 0
```

This means that the error occurred in line 5 of `Ginger`, and the values of the variables at that point were as given. We deduce that `Ginger` was trying to compute 0^{-2}:

```
>   0^(-2);
```

```
Error, numeric exception: division by zero
```

Remark. The punctuation of the above procedures appears to omit some closing semicolons (;) or colons (:). This is deliberate: The last statement in a procedure does not need a semicolon, and likewise neither the last statement in a `for`-loop nor the last clause in an if-statement needs a terminator. I leave these off when I can, not out of laziness but rather because inserting a statement *after* such a terminating statement can have a larger effect than just the execution of that new statement. The value returned by a for-loop is the value of the last statement executed; likewise, the value returned by a procedure is usually the value of the last statement executed. Adding another statement will change that, and if I didn't mean that to happen I want an error message. This is merely a personal use of this punctuation feature. You may, if you like, always put terminators on your

Maple statements; it will make no difference to the program as written. I have been told that my practice goes quite against what is usually taught in computer science courses, namely, that as much as possible, code should be written so that the insertion of a valid statement either before or after any other valid statement should not generate a syntax error. Please use what works for you.

3.1.1 Structured Types

Maple has types, as you have seen from the previous examples. The types are dynamically verified on execution of a given piece of code. The basic types correspond to the data structures mentioned in Chapter 1. There is a mechanism for querying the type of an expression, the `type` command. This is especially useful for recognizing complicated "structured" types. For example, the expression x^2 is of type `algebraic`, but it is also of type `anything^integer`.

```
>   restart;
>   f := 1 + (x+z)^3 + tan((x+z)^3)*sin(y);
```
$$f := 1 + (x + z)^3 + \tan((x + z)^3)\sin(y)$$
```
>   whattype( f );
```
$$+$$
```
>   type( f, '+' );
```
$$true$$
```
>   type( f, algebraic );
```
$$true$$
```
>   type( x^2, algebraic );
```
$$true$$
```
>   type( x^2, anything );
```
$$true$$
```
>   type( x^2, set );
```
$$false$$

Type-checking is not tied in to the assume facility, so assumptions about the names are not checked: the following is a syntactic check to see whether the input is of type "complex," not an inquiry into whether the input is a complex-valued function of its possibly complex variables:

```
>   type( x^2, complex );
```
$$false$$

You can construct complicated types out of basic types:

```
>   type( x^2, name^posint );
```
$$true$$

Type-checking is one of the simplest and most useful ways of preventing bugs in your program and of tracking down the ones that occur. If you do not do your own argument type-checking, you run the risk of getting very cryptic error messages on occasion.

```
>   b := proc( x::integer ) 2*x end proc:
>   b( Pi );
```

```
Error, invalid input: b expects its 1st argument, x, to be of
type integer, but received Pi
```

```
>   b( -1 );
```

$$-2$$

Exercises

1. Write a procedure that takes one argument and squares it if it is bigger than 1.

2. Try `subs(17=15, eval(Fred))` and see what you get. This trick is not recommended, because sometimes the results are surprising. Try `subs(1=2, eval(Ginger))` and comment.

3.1.2 Example: Modified Gram–Schmidt

The following example shows, by a series of versions of the same program, how to program with loops, how to use the Matrix constructors and vector indexing, and the benefits of specializing code for specific data types. We begin with a brief review of the Modified Gram–Schmidt (MGS) algorithm for computing an orthonormal factoring of a matrix.

A rectangular m-by-n matrix A may be factored into a product of an m-by-m orthonormal matrix Q and an upper-triangular matrix R, as follows:

$$A = QR .$$

The classical Gram–Schmidt process for computing Q and R is known to be numerically unstable, in that for a matrix A with floating-point entries the resulting matrix Q does not always nearly satisfy $Q^T Q = I$. A simple modification (interchanging the order of the loops), known as Modified Gram–Schmidt, stabilizes the process. See [31] for the algorithm and its analysis. That analysis embeds MGS into Householder QR factoring. Therefore, since the LinearAlgebra package in Maple already has QR factoring (using the NAG routines), there is no need to write this program in Maple except as an example. See ?QRDecomposition.

In Figure 3.1 we see the first implementation of this process, which can also be taken as a description of the algorithm. This implementation shows how to use several simple Maple programming features.

```
MGS1 := proc( iA :: Matrix )
   local Q, R, i, j, k, m, n;
   description "Modified Gram-Schmidt using for-loops.";
   use LinearAlgebra in
      m := RowDimension( iA );
      n := ColumnDimension( iA );
      R := Matrix( n, n, shape=triangular[upper] );
      Q := Matrix( m, n, iA );
      for k to n do
         R[k,k] := ( add(conjugate(Q[i,k])*Q[i,k], i=1..m) )^(1/2);
         for i to m do
            Q[i,k] := Q[i,k]/R[k,k]
         end do;
         for j from k+1 to n do
            R[k,j] := add( conjugate(Q[i,k])*Q[i,j], i=1..m );
            for i to m do
               Q[i,j] := Q[i,j] - R[k,j]*Q[i,k]
            end do
         end do
      end do
   end use;
   (Q, R)
end proc:
```

Figure 3.1: A Maple program that uses for loops in MGS

The line

```
    MGS1 := proc( iA :: Matrix )
```

starts the procedure, names it MGS1, and says that the procedure expects one
input argument, which it will refer to as iA, and that this argument must be of
type Matrix.

The line

```
    local Q, R, i, j, k, m, n;
```

declares the local variables used in this procedure (see Section 3.4).
The line

```
    description "Modified Gram-Schmidt using for-loops.";
```

is the only kind of comment that is kept in the procedure body when it is loaded
into Maple. It is useful to put description strings on many procedures. Here this
helps to distinguish several versions one from the other.

The lines

```
        use LinearAlgebra in
           m := RowDimension( iA );
           n := ColumnDimension( iA );
```

allow access to the LinearAlgebra routines (here RowDimension and Column-
Dimension) in their short name format without altering access to the LinearAlgebra package explicitly, outside the scope of the "use...end use" construct.
It is generally a good idea not to use with inside a procedure: let the user specify
exactly the packages needed, and only those needed.

The lines

```
R := Matrix( n, n, shape=triangular[upper] );
Q := Matrix( m, n, iA );
```

declare and initialize the matrices used in the factoring. This makes a copy of the
input matrix iA (needed so that we do not overwrite the input). The use of the
shape= construct saves storage; the matrix R will be upper triangular, and there
is no need to store the zeros below the diagonal.

The lines

```
for k to n do
    R[k,k] := ( add(conjugate(Q[i,k])*Q[i,k],
                    i=1..m) )^(1/2);
    for i to m do
        Q[i,k] := Q[i,k]/R[k,k]
    end do;
```

start a loop over the columns of Q. The first statement in the loop computes the
2-norm of the kth column of Q, and the nested loop over the row index i scales
the kth column so that its 2-norm is 1.

The lines

```
for j from k+1 to n do
    R[k,j] := add( conjugate(Q[i,k])*Q[i,j], i=1..m );
    for i to m do
        Q[i,j] := Q[i,j] - R[k,j]*Q[i,k]
    end do
end do
```

start a nested loop over the columns to the right of the kth column, replacing the
jth column by a vector orthogonal to the kth column, for each $j = k + 1, k +
2, \ldots, n$. Inside this loop on j is an explicit loop on the row index i to replace
each element of the jth column of Q with the appropriate multiple computed
from the dot product. The add construct is itself a disguised loop over the row
index, $1 \le i \le m$.

The final lines

```
    end do
  end use;
  (Q, R)
end proc:
```

close off the loop over k, end the use environment within which we could use
LinearAlgebra routines, return the expression sequence Q, R to the user or

```
MGS3 := proc( iA :: Matrix )
   local Q, R, j, k, m, n;
   description "Modified Gram-Schmidt using block indexing and datatyping.";
   use LinearAlgebra in
      m := RowDimension( iA );
      n := ColumnDimension( iA );
      R := Matrix( n, n, shape=triangular[upper], datatype=float[8] );
      Q := Matrix( m, n, iA, datatype=float[8] );
      for k to n do
         R[k,k]  := Norm( Q[1..m,k], 2 );
         Q[1..m,k]  := Q[1..m,k]/R[k,k];
         R[k,k+1..n] := HermitianTranspose(Q[1..m,k]) . Q[1..m,k+1..n];
         Q[1..m,k+1..n] := Q[1..m,k+1..n] - Q[1..m,k].R[k,k+1..n];
      end do
   end use;
   (Q, R)
end proc:
```

Figure 3.2: A Maple program that uses block indexing in MGS

calling program, and end the procedure. At this point, if the input A contained floating-point entries, then $A = QR$ approximately, and $Q^T Q = I$ approximately. Now, of course, if the input A had been exact or symbolic (say a matrix of integers), then Q and R would contain exact square roots of integers. Sometimes this is what is wanted, but the computation can be expensive in that case and the results cumbersome. See the exercises.

In Figure 3.2 we see a more developed (and therefore simpler) version of the same program. It loops only over the columns; each of the nested loops in the original program has been replaced by a vector or block matrix operation. For those who are used to MATLAB this is a familiar thing to do: It improves the readability and maintainability of programs, and in the case of MATLAB, greatly enhances the efficiency of the programs.

The statement

```
Q[1..m,k]    := Q[1..m,k]/R[k,k];
```

replaces the kth column of Q with one scaled so that its 2-norm is 1. The line

```
R[k,k+1..n]  := HermitianTranspose(Q[1..m,k])
                 . Q[1..m,k+1..n];
```

computes the dot product of the current column of Q with all the remaining columns at once; the result is a row vector that we store in the kth row of R. The line

```
Q[1..m,k+1..n]  := Q[1..m,k+1..n]
                    - Q[1..m,k] . R[k,k+1..n];
```

replaces (all at once) each higher-numbered column with a vector that has had its component in the direction of the kth column of Q removed.

Surprisingly, if we had failed to specify that the data type of the matrices is float[8] or complex[8], this program would run slightly *slower* than the loop version. The reasons for that include the fact that if we had not specialized the data type of the matrices, Maple would have to do continual type-checking during the computation. Another speed factor is that we are not using the "programmer entry points" for the LinearAlgebra routines, and thus even more type-checking is done inside the loops. We can improve the speed by a factor of about 3 by adding the line datatype=float[8] to the Matrix constructions of Q and R, as done in the figure.

I called the version of the program that did not use these declarations MGS2 (not shown), and we see below that MGS3 is fastest, followed by MGS1 and then, just last, MGS2.

```
>    restart;
>    read "D:/books/ess/programs/mgs.mpl";
>    with(LinearAlgebra):
>    m, n := 100,50;
```
$$m, n := 100, 50$$
```
>    A := RandomMatrix(m,n);
```
$$A := \left[100 \text{ x } 50 \text{ Matrix } \text{ Data Type}: \text{ anything} \right.$$
$$\left. \text{Storage}: \text{ rectangular } \text{ Order}: \text{ Fortran\_order} \right]$$
```
>    Digits := trunc(evalhf(Digits));
```
$$Digits := 14$$
```
>    st := time(): Q,R := MGS1( evalf(A) ): time()-st;
```
$$23.994$$
```
>    Norm( Q.R - A, infinity )/Norm(A,infinity);
```
$$.46274924080626 \; 10^{-13}$$
```
>    Norm( HermitianTranspose(Q).Q
>          - IdentityMatrix(n), infinity );
```
$$.954187466886691205 \; 10^{-12}$$
```
>    st := time(): Q,R := MGS2( evalf(A) ): time()-st;
```
$$28.244$$
```
>    Norm( Q.R - A, infinity)/Norm(A,infinity);
```
$$.11202527542813 \; 10^{-13}$$
```
>    Norm( HermitianTranspose(Q).Q
>          - IdentityMatrix(n), infinity );
```
$$.3428546509198877102 \; 10^{-12}$$

```
>   st := time(): Q,R := MGS3( evalf(A) ): time()-st;
```
$$11.619$$

So we see that this version is faster than the original, or the modification.

```
>   Norm( Q.R - A, infinity)/Norm(A,infinity);
```
$$.11202527542813 \ 10^{-13}$$
```
>   Norm( HermitianTranspose(Q).Q
>          - IdentityMatrix(n), infinity );
```
$$.3428546509198771102 \ 10^{-12}$$

Exercises

1. Try the programs out with a small matrix that has integer entries (that is, really integers, like 2, and not floating-point integers like 2.0). The results should contain square roots and be somewhat cumbersome. Compare the answers and the speed with which you get them if you use `evalf` on the input matrix first.

2. Redesign the $A = QR$ factoring so that no square roots are taken. This means that $Q^T Q = D$ would be diagonal but not necessarily the identity matrix. Show how to use this factoring to solve $Ax = b$ where A and b contain only rational entries. Explain why avoiding square roots is a good idea in a symbolic system.

3. Try the built-in `LinearAlgebra` routine `QRDecomposition` on a random 100-by-50 matrix (be careful to use `evalf` to make it a matrix of floating-point numbers). On my machine it takes 0.2 seconds, whereas MGS3 above took 11 seconds. This is why the example of this section is just an example; there is no need to implement MGS in Maple.

3.2   Operators and Modules

Operators are functions from one abstract space to another. The mathematical usage of the word "operator" is usually reserved for functions on functions. In Maple this is not always the case, and indeed a Maple operator is usually just a function. But there are important mathematical operators in Maple. For example, consider the differentiation operator $D : f \rightarrow f'$. The Maple notation for this operator is just D. Historically, D was introduced into Maple to solve one particular problem: to find a way to represent the value of the derivative of an unknown function at a particular point (the natural command `subs(x=3, diff(f(x), x))` doesn't work; it produces `diff(f(3),3)` which gives an error). This is represented in Maple now as `D(f)(a)`, meaning $f'(a)$. I remark that Newton's notations $\dot{x}(a)$ and $x'(a)$ survive today because they express this idea concisely; contrariwise, the

Leibniz notation df/dx survives because it has its own, algebraic and mnemonic, uses. The notation

$$\frac{df}{dx}\bigg|_{x=a}$$

is relatively cumbersome. At the time of the first edition of this book I knew of only one computer algebra system (that of the HP48 series calculators) that used this notation, but it is now available in Maple (for explicit constants a) by use of the eval command:

```
>  restart;
>  eval( diff(f(x),x), x=17 );
```

$$\left(\tfrac{\partial}{\partial x}\, f(x)\right)\bigg|\, x = 17$$

Speaking personally, I do not like that notation, although it does have a temporal advantage for teaching purposes: It is clearer with this notation that you differentiate first, then evaluate at $x = a$.

Operators in Maple are represented by "arrow" notation:

```
>  restart;
>  f := t -> t*sin(t) ;
```

$$f := t \to t \sin(t)$$

We may now evaluate this function by applying it to any input arguments that we please.

```
>  f(0);
```

$$0$$

```
>  f(x^2+a);
```

$$\left(x^2 + a\right) \sin\left(x^2 + a\right)$$

Those were examples of function *application*. We *applied* the operator f to the arguments 0 and $x^2 + a$. That result, $\left(x^2 + a\right) \sin\left(x^2 + a\right)$, is the result of *composing* the operator f with the operator $h : t \mapsto t^2 + a$ and *applying* the resulting operator to the argument x. *Application is not the same as composition.*

Application versus composition

The following rules of thumb do not give the whole story, but they help.

- When you compose two operators, the result is an operator. The domain and range of the two operators must be compatible.

- When you apply an operator to an argument, the result is an expression. The argument must be in the domain of the operator.

Mathematically, this is a difference in *what you think about the answer*, and is subject to a little deliberate overlap when this is useful. In Maple the distinction is usually enjoined by syntax.

The Maple syntax for function application uses parentheses ( ). The function f applied to the argument u is written `f(u)`.

The composition operator in Maple is the "at"-sign @. This is the closest ASCII symbol available to the standard mathematical notation ∘.

The distinction between application and composition will be made clearer with further examples, but note for now that the result of composing f with g is the operator $f \circ g : t \mapsto f(g(t))$, whereas the application of this operator to the argument x gives the *expression* $f(g(x))$. To be concrete, if $f = t \to \sin(t + \phi)$ and $g = u \to \exp(u)$, then $f \circ g = z \to \sin(\exp(z) + \phi)$ is an operator, while $f(g(x)) = \sin(\exp(x) + \phi)$ is an expression, because we think of x as being a real number.

We differentiate operators with D, not `diff`:

> `D(f);`

$$t \to \sin(t) + t \cos(t)$$

More operator examples

One can apply *type-checking* to operators, as follows (it is the same as for general procedures):

> `g := (t::complex(numeric)) -> t*arcsin(t) ;`

$$g := t::\text{complex}(\textit{numeric}) \to t \arcsin(t)$$

> `g(0);`

$$0$$

> `g(x);`

`Error, invalid input: g expects its 1st argument, t, to be of type complex(numeric), but received x`

> `g(1+I);`

$$(1 + I) \arcsin(1 + I)$$

> `g(1.+I);`

$$-.3950356295 + 1.727514494\,I$$

Note that the elementary functions in Maple work over the complex plane, with good closures for functions with branch cuts, as evidenced by the following calls (note the use of signed zero in the imaginary part to indicate which side of the branch we are on). See Appendix A.

> `g( 2. );`

$$3.141592654 - 2.633915794\,I$$

```
>  g( 2.+0.*I );
```
$$3.141592654 + 2.633915794\,I$$
```
>  g( 2.-0.*I );
```
$$3.141592654 - 2.633915794\,I$$

The following is one definition of an antidifferentiation operator. Typically, this is denoted by I, but that symbol is taken in Maple. If we don't want to rebind I using `interface(imaginaryunit)`, then we can be creative. MATLAB uses eye, so perhaps...

```
>  restart;
>  Aye := f -> unapply(int(f(x),x),x);
```
$$Aye := f \rightarrow \mathrm{unapply}(\int f(x)\,dx,\ x)$$
```
>  Aye( t->t^2 );
```
$$x \rightarrow \frac{1}{3}x^3$$

Some simplifications of operators are automatic:
```
>  Aye( t->1/t );
```
$$\ln$$
```
>  t -> sin(t);
```
$$\sin$$

That is just the `sin` operator, and the `ln` operator appeared automatically on integration of the $t \rightarrow 1/t$ operator (which doesn't have a name). One of the most confusing aspects of automatic simplification of operators is that *constants are treated as operators*.

```
>  x -> 4;
```
$$4$$
```
>  4(3);
```
$$4$$
```
>  2(anything);
```
$$2$$

Probably the most common error with operators is to leave the multiplication symbol * out of the input to a parenthesized expression, which changes the meaning from multiplication to function application: application of a constant operator.

```
>  2(a+b);
```
$$2$$

That result is correct, and this feature is often used in Maple. It takes some getting used to, though, and provides a good trap for the unwary.

Operators are sometimes used as notation in expressions, as in this series computation.

```
>  restart;
>  series( f(x), x=a, 4 );
```

$$f(a) + D(f)(a)(x-a) + \frac{1}{2}\left(D^{(2)}\right)(f)(a)(x-a)^2$$
$$+ \frac{1}{6}\left(D^{(3)}\right)(f)(a)(x-a)^3 + O\left((x-a)^4\right)$$

[Notice that this assumes enough smoothness of the unknown function f.] Again, application is not the same as composition. Here we have applied the operator D to the operator f, which yields another operator $D(f)$. This is best thought of as application, not composition, because the operator f is *in of the domain of D*, and thus is best thought of as an *argument*.

We then *apply* the resulting operator $D(f)$ to a to get $D(f)(a)$, or $f'(a)$. We *compose* D with itself, and apply the resulting operator to f to get $(D \circ D)(f) = D^{(2)}(f)$, and apply this resulting operator to the argument a to get $D^{(2)}(f)(a) = f''(a)$. If we confuse application with composition, we may generate errors in Maple, or simply not get the result we intend.

Maple knows the chain rule (and assumes that all functions are nice enough that it is universally true):

```
>  D( g@h );
```

$$((D(g))@h)\, D(h)$$

Maple also knows the product rule, and again assumes that all functions are nice enough that it is universally true:

```
>  D( g*h );
```

$$D(g)\, h + g\, D(h)$$

If you know that h is constant and x is the independent variable, you can use the remember table of D to simplify some computations:

```
>  D(h) := 0;
```

$$D(h) := 0$$

```
>  D(x) := 1;
```

$$D(x) := 1$$

That told Maple that $dx/dx = 1$.

```
>  D( F@x );
```

$$(D(F))@x$$

```
>  D( x^2 );
```

$$2\,x$$

In contrast, if you differentiate an arbitrary operator squared, you get a derivative hanging around from the chain rule:

```
>  D( g^2 );
```

$$2 D(g) g$$

3.2.1 A Module for Finite-Difference Operators

Operators can act on functions (other operators) or on numerical objects. As an example, we investigate the finite-difference operators. Since there are several related finite-difference operators, we make a small *module*. [As stated before, a module in Maple is a programming construct, and not a module from algebra.] Since the difference (step) is usually common to all related operators, we make this a *parameterized module*, by putting the module inside a procedure that gets called with the step size h. The module then inherits the step size by nested lexical scopes. See Section 3.4.4 for more discussion of scoping rules.

```
>  restart;
```

Because we would like to use I (and not Aye) for the integration operator, we free up the binding of I to the imaginary unit:

```
>  interface(imaginaryunit=j);
```

You may choose instead not to clobber the useful letter j, and instead use something more arcane, like `'\sqrt{-1}'`. Now we present the procedure that generates the parameterized module:

```
>  finite_differences := proc( h::{name,complex(numeric)} )
>      description "Generator for finite difference"
>              " operators with step size h.";
>      module()
>      export I, E, Delta, delta, S;
>      local t, x, k, N;
>        E := f -> (x->f(x+h));
>        Delta := f -> (t->((E(f)(t)-f(t))/h));
>        delta := f -> (x -> (f(x+h)-f(x-h))/(2*h) );
>        I := f -> unapply( int(f(t), t=0..x), x );
>        S := f -> unapply( sum(f(k*h), k=1..N), N );
>      end module;
>  end proc:
```

Remarks

1. The procedure `finite_differences` takes as argument a variable that is called h inside the procedure. As usual the actual parameter might be something else.

2. Note the use of the *structured type* in the type-checking:

   ```
   h::{name,complex(numeric)}
   ```

means that h can be either a name or a complex number (which includes real numbers). It is quite common to look at differences with step $h = 1$, for example.

3. The use of "unapply" in I and S means that when the resulting operators are printed, the function that they were called with will appear explicitly in the procedure body.

4. Earlier I used `Aye` for the integration operator, and here I use `I`, which I prefer. Both ways are shown, for information.

Every procedure returns a result (which might, it is true, be a NULL result). By default, the last object computed in the procedure is returned. In this example, the object happens to be a *module*. See `?module`. The purpose of a module is to gather together related routines and data structures, allowing them to share information amongst each other without conflicting with objects in other routines or at the top level.

A module makes certain of its internal objects visible to the user or to other programs. These are called its *exports*. Here the exports are the operators E, Δ, δ, and I.

To start using the module, we must fix a parameter by calling the procedure with a symbol or number, here ε, as the name for the step size.

```
>    hop := finite_differences( epsilon );
```

$hop :=$ **module**$()$ **local** t, x, k, N; **export** I, E, Δ, δ, S; **end module**

The module was instantiated with the command above. Now we use the `with` command to bind the names of the exports at the global level, to make the short forms available for interactive use:

```
>    with( hop );
```

$$[\Delta, E, I, S, \delta]$$

Note that Maple sorted the list of exports (capital letters come first, which is strange to people who don't use Unix).

```
>    esin := E( sin );
```

$$esin := x \rightarrow \sin(x + \varepsilon)$$

The result above is an *operator*. It can be *applied* to an argument, say ϕ:

```
>    esin( phi );
```

$$\sin(\phi + \varepsilon)$$

Since it is an operator, we can shift it, too:

```
>    E( esin );
```

$$x \rightarrow esin(x + \varepsilon)$$

We can apply that operator to an argument:

```
>  E( esin )(a);
```
$$\sin(a + 2\,\varepsilon)$$

I remind you that functional composition is denoted by the @ symbol, which is the closest that ASCII comes to the mathematical symbol ∘. Repeated composition uses @@, in analogy with the FORTRAN use of ** to denote exponentiation. In Maple, the precedence of @@ is such that you must use parentheses to indicate the desired behaviour.

```
>  (E@@2)(sin);
```
$$x \to (x \to \sin(x + \varepsilon))(x + \varepsilon)$$
```
>  (E@@3)(sin)(a);
```
$$\sin(a + 3\,\varepsilon)$$

We now look at some integration, differentiation, and summation examples:

```
>  F := I( sin );
```
$$F := x \to -\cos(x) + 1$$
```
>  D( F );
```
$$\sin$$
```
>  DeltaF := Delta( F );
```
$$DeltaF := t \to \frac{E(F)(t) - F(t)}{\varepsilon}$$

Forward difference:
```
>  DelF := DeltaF( t );
```
$$DelF := \frac{-\cos(t + \varepsilon) + \cos(t)}{\varepsilon}$$
```
>  limit( DelF, epsilon = 0 );
```
$$\sin(t)$$

Central difference:
```
>  DelF := delta( F )(xi);
```
$$DelF := \frac{1}{2}\frac{-\cos(\xi + \varepsilon) + \cos(-\xi + \varepsilon)}{\varepsilon}$$
```
>  limit( DelF, epsilon=0 );
```
$$\sin(\xi)$$

The following evaluates the antidifference (i.e. sum) $\sum_m (\varepsilon m)^2$.
```
>  S( m->m^2 );
```
$$N \to \frac{1}{3}\varepsilon^2(N+1)^3 - \frac{1}{2}(N+1)^2\varepsilon^2 + \frac{1}{6}\varepsilon^2(N+1)$$

```
>   sinsum := S( n->sin(n*Pi) );
```

$$sinsum := N \rightarrow -\frac{1}{2}\sin((N+1)\,\varepsilon\,\pi) + \frac{\frac{1}{2}\sin(\varepsilon\,\pi)\cos((N+1)\,\varepsilon\,\pi)}{\cos(\varepsilon\,\pi)-1}$$
$$+\frac{1}{2}\sin(\varepsilon\,\pi) - \frac{1}{2}\frac{\sin(\varepsilon\,\pi)\cos(\varepsilon\,\pi)}{\cos(\varepsilon\,\pi)-1}$$

```
>   sinsum( m );
```

$$-\frac{1}{2}\sin((m+1)\,\varepsilon\,\pi) + \frac{\frac{1}{2}\sin(\varepsilon\,\pi)\cos((m+1)\,\varepsilon\,\pi)}{\cos(\varepsilon\,\pi)-1} + \frac{1}{2}\sin(\varepsilon\,\pi)$$
$$-\frac{1}{2}\frac{\sin(\varepsilon\,\pi)\cos(\varepsilon\,\pi)}{\cos(\varepsilon\,\pi)-1}$$

```
>   combine( %, trig );
```

$$\frac{-\sin(\varepsilon\,\pi\,m) + \sin(\varepsilon\,\pi\,m + \varepsilon\,\pi) - \sin(\varepsilon\,\pi)}{2\cos(\varepsilon\,\pi)-2}$$

```
>   eval( %, epsilon=1/m );
```

$$-2\frac{\sin(\dfrac{\pi}{m})}{2\cos(\dfrac{\pi}{m})-2}$$

```
>   limit( %%, epsilon = 0 );
```

$$0$$

3.2.2   Remarks on Mathematical Operators

Mathematical operators such as the shift operator E defined above, the finite divided difference operators, the differentiation operator D, the "kernel" operators $K : f \rightarrow \int_a^b k(\cdot, x)f(x)\,dx$, and many others, provide useful constructs in applied mathematics. By artful (ab)use of notation, useful relations between some of the operators can be exploited.

$$E(f)(x) = f(x+h) = f(x) + D(f)(x)h + \frac{1}{2!}(D \circ D)(f)(x)h^2 + \cdots$$
$$= (I + hD + \frac{1}{2!}h^2 D^{(2)} + \cdots)(f)(x)$$
$$= \exp(hD)(f)(x),$$

and so $E = \exp(hD)$ where we interpret the exponential of the operator hD as the Taylor series of exp with the *powers* replaced by *repeated composition*, so $D^{(2)}$ is interpreted as $D \circ D$, and thus $D^{(2)}(f)(a) = f''(a)$. Note that the leading 1 in the Taylor series for exp is replaced by the *identity operator* $I : x \rightarrow x$.

Exercises

1. Write an "arithmetic–geometric mean" operator that takes two numbers a and b and returns $(a+b)/2$ and $(ab)^{1/2}$. [Note: Maple's square root function sqrt does some processing before simplifying its input α to $\alpha^{1/2}$. You can save some execution time, not crucial here but perhaps of interest, by directly using $\alpha^{1/2}$.] My operator for this procedure is only 27 characters long, not counting spaces but including the semicolon; is there a shorter way? See also ?GaussAGM.

2. Write a procedure `convert/polyop` that will convert a polynomial in a variable (e.g., $p = 3 + d + \frac{1}{2}d^2$) to the analogous mathematical operator (e.g., $p = 3I + d + \frac{1}{2}d \circ d$). Be careful of the constant term.

3. Write a second order difference operator that approximates $f''(x)$ by $(f(x+h) - 2f(x) + f(x-h))/h^2$.

4. Show that the forward divided difference operator Δ that takes $x \to f(x)$ to $x \to (f(x+h) - f(x))/h$ is related to the differentiation operator D by $\Delta = (\exp(hD) - 1)/h$. Invert this relationship and use power series in Maple to get a 12th-order accurate finite difference approximation to D.

5. Find a similar relationship between δ and D, invert it, and find an 8th-order finite difference approximation to D using central differences.

6. Using `CompanionMatrix`, `lcoeff`, `Eigenvalues`, and `evalf`, write an operator that finds all roots of a given polynomial with numerical coefficients.

3.3   Data Structures

What follows is a very brief overview of Maple's data structures. For a more in-depth look, see [28], and for complete details consult [44]. There are many built-in data structures in Maple, often quite different from those of FORTRAN, C, or Pascal. The main data structures are

1. algebraics

2. lists

3. sets

4. unevaluated or inert function calls

5. tables

6. Records (see ?Record), which correspond to C structs,

7. relations

8. series

9. strings

10. indexed names

11. sequences

and the numeric data structures for integers, fractions, floats, and complex numbers.

An "algebraic" data structure (indeed, every data structure in Maple) is represented internally as a Directed Acyclic Graph, or DAG for short. For most users it is best to think of an algebraic data structure as simply an expression containing symbols. For example, the expression

```
1 + (x+z)^3 + tan((x+z)^3)*sin(y)
```

is of algebraic type. It is explored in the session below, and a simplified version of the Maple DAG is sketched in Figure 3.3. I remark that the session below is sensitive to *ordering changes*; Maple orders subexpressions in a session-dependent fashion, because Maple uses *address order* for efficiency. This can cause confusion when execution of a Maple script produces different results from previous executions.

```
>  restart;
>  f := 1 + (x+z)^3 + tan((x+z)^3)*sin(y);
```
$$f := 1 + (x + z)^3 + \tan\left((x + z)^3\right)\sin(y)$$
```
>  nops(f);
```
$$3$$

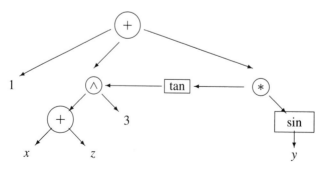

Figure 3.3: A simplified DAG of $1 + (x + z)^3 + \tan((x + z)^3)\sin(y)$

```
>   op(1,f);
```
$$1$$

```
>   op(2,f);
```
$$(x + z)^3$$

```
>   op(3,f);
```
$$\tan\left((x + z)^3\right) \sin(y)$$

```
>   op(1,op(2,f));
```
$$x + z$$

The op command is a useful "low-level" procedure for picking apart the operands of an object. The command nops counts the number of operands in an object.

Function calls use parentheses, such as sin(x) or ChebyshevForm(x, A, 12). We will see uses for unevaluated function calls as data structures later.

Tables use parentheses in their creation, as in T := table() or A := array(0..3), and square brackets to reference individual entries, as in A[3].

Sequences consist of two or more objects separated by commas.

```
>   restart;
>   e_seq := 1,2,3,4;
```
$$e\_seq := 1, 2, 3, 4$$

Lists are simply expression sequences in square brackets. The ith element of an expression sequence or list can be selected as though it were in an array.

```
>   L := [1,2,3,4];
```
$$L := [1, 2, 3, 4]$$

```
>   M := [4,3,2,1];
```
$$M := [4, 3, 2, 1]$$

The lists L and M above are different, since order is important for lists. To select the third entry in each, use e_seq[3], L[3], and M[3].

```
>   e_seq[3];
```
$$3$$

```
>   L[3];
```
$$3$$

```
>   M[3];
```
$$2$$

Sets, on the other hand, use braces (curly brackets), and order is unimportant:

```
>   S := {1, 2, 3, 4}:
>   T := {2, 3, 4, 1}:
```

In this example, S and T define *the same set*, and indeed Maple will detect this (by hashing) and store only one copy of this set. This can be verified by using the addressof function to locate the data structures in memory.

```
>   addressof( S );
```

$$4392076$$

```
>   addressof( T );
```

$$4392076$$

The series data structure arises only from a call to the series, taylor, or asympt commands. Manipulate these by further calls to those commands, or by converting them to sums of terms. The series data structure is a *sparse* data structure: terms with zero coefficients are not stored. For details, see ?type[series].

```
>   s := series( sin(x), x );
```

$$s := x - \frac{1}{6}x^3 + \frac{1}{120}x^5 + O(x^6)$$

```
>   c := series( cos(x), x );
```

$$c := 1 - \frac{1}{2}x^2 + \frac{1}{24}x^4 + O(x^6)$$

```
>   s+c;
```

$$\left(x - \frac{1}{6}x^3 + \frac{1}{120}x^5 + O(x^6) \right)$$
$$+ \left(1 - \frac{1}{2}x^2 + \frac{1}{24}x^4 + O(x^6) \right)$$

```
>   whattype( s+c );
```

$$+$$

```
>   sumo := series( s+c, x );
```

$$sumo := 1 + x - \frac{1}{2}x^2 - \frac{1}{6}x^3 + \frac{1}{24}x^4 + \frac{1}{120}x^5 + O(x^6)$$

```
>   whattype( sumo );
```

$$series$$

```
>   convert( sumo, polynom );
```

$$1 + x - \frac{1}{2}x^2 - \frac{1}{6}x^3 + \frac{1}{24}x^4 + \frac{1}{120}x^5$$

```
>   whattype( % );
```

$$+$$

Sometimes the result from series is not of type series. This occurs particularly when the result is a Puiseux series.

```
>    series( sin(x^(1/3)), x, 3 );
```

$$x^{(1/3)} - \frac{1}{6}x + \frac{1}{120}x^{(5/3)} - \frac{1}{5040}x^{(7/3)} + O(x^3)$$

```
>    whattype( % );
```

$$+$$

```
>    lprint( %% );
```

```
x^(1/3)-1/6*x+1/120*x^(5/3)-1/5040*x^(7/3)+O(x^3)
```

```
>    series( sin(x), x);
```

$$x - \frac{1}{6}x^3 + \frac{1}{120}x^5 + O(x^6)$$

```
>    lprint( % );
```

```
series(1*x-1/6*x^3+1/120*x^5+O(x^6),x,6)
```

Strings are reasonably important in Maple. They are used mainly for messages, labels, text, and to access file names in the read command. See ?string.

Names are either of type symbol or of type indexed. A symbol can be either a simple sequence of one or more alphanumeric characters (starting with an alphabetical character), which is called a simple symbol or simple name, or a symbol can be any sequence of alphanumeric characters including spaces or punctuation, all enclosed in backquotes.

```
>    restart;
>    x;
```

$$x$$

```
>    xab12bang_this_is_a_simple_name;
```

$$xab12bang\_this\_is\_a\_simple\_name$$

```
>    'This is still a simple name: it just has
>    weird characters in it@#&*:-)';
```

This is still a simple name : it just has weird characters in it@#&∗ : −)

I typed that long name all on one line, and Maple's Export-to-LaTeX feature broke the line without putting an escape or newline into it. To reproduce the output you see above, type the long name all on one line in a worksheet.

Indexed names are objects that look like A[3] or Database[1, 2, sock_drawer], i.e., tables or Matrices or Vectors or Arrays. You need not explicitly create a table before you start using indices. Maple will automatically create a table if it has to. Indexed names are used most often for Matrices and Vectors. To ensure that Maple creates the right sort of table, an explicit creation before assignment is recommended.

```
>    restart;
>    x := T[3];
```

$$x := T_3$$

```
>   eval( T );
```

$$T$$

```
>   T[5] := 7;
```

$$T_5 := 7$$

```
>   eval( T );
```

$$\text{table}([5 = 7])$$

Maple did not create a table for T until we assigned something to T[5].

```
>   A := Matrix( 3, 3, (i,j)->i^(j-1) );
```

$$A := \begin{bmatrix} 1 & 1 & 1 \\ 1 & 2 & 4 \\ 1 & 3 & 9 \end{bmatrix}$$

```
>   lprint( A );
```

```
Matrix(3,3,{(1, 1) = 1, (1, 2) = 1, (1, 3) = 1, (2, 1) = 1,
(2, 2) = 2, (2, 3) = 4, (3, 1) = 1, (3, 2) = 3, (3, 3) = 9},
datatype = anything,storage = rectangular,
order = Fortran_order,shape = [])
```

```
>   A[1,2];
```

$$1$$

```
>   B := Vector( 2, i->b[i] );
```

$$B := \begin{bmatrix} b_1 \\ b_2 \end{bmatrix}$$

```
>   C := Matrix( 2, 2, (i,j)->c[i,j] );
```

$$C := \begin{bmatrix} c_{1,1} & c_{1,2} \\ c_{2,1} & c_{2,2} \end{bmatrix}$$

```
>   adj := LinearAlgebra:-Adjoint( C );
```

$$adj := \begin{bmatrix} c_{2,2} & -c_{1,2} \\ -c_{2,1} & c_{1,1} \end{bmatrix}$$

```
>   dt := LinearAlgebra:-Determinant( C );
```

$$dt := c_{1,1}\,c_{2,2} - c_{1,2}\,c_{2,1}$$

Notice that the answer contains indexed names. Creation of a Matrix lets us access its entries by indexing.

The numeric data structures (integer, fraction, rational, float, hardware float, and complex numeric) are largely transparent in their uses. Floats are usually contagious, which means that if one element of a numeric structure is a float, then automatically the whole structure is converted to floating point, if possible.

Hardware floats are automatically converted to Maple floats on return from evalhf, unless they are kept in a Vector or Matrix or hfarray (this last is now deprecated, and will be replaced in future versions of Maple).

3.4 Local versus Global versus Environment Variables

Local variables are local to a procedure or module, are usually available only to subroutines defined internally by that procedure or module, and disappear once that procedure or module completes its execution (unless they are "exported"). Using local variables is an excellent way for hiding information that is irrelevant to other procedures, and for freeing up short names to use elsewhere. In the procedure `Ginger` above, the variables `i,j,k`, and `s` are all local, and that example shows how to declare local variables in procedures. You cannot declare local variables in an operator. Variables local to a module are available to all subprocedures of that module. A good example of this can be found in Figure 2.1 in Chapter 2. There, the procedure `auxiliary` is local to the module, but is accessible to the procedure `hide`, which is made available to the user.

3.4.1 Exporting Local Variables

Sometimes it is useful to export local variables, which are unique (although two or more different ones may look the same) and difficult to "touch." See the chapter "When local variables leave home" of [44].

An amusing demonstration of exporting local names was given on USENET in 1993 by Frederic W. Chapman, along the following lines. [This example presumes some knowledge of *Star Trek: The Next Generation*, but hopefully the point will still get across even if the reader is not a Trekker.]

```
>   restart;
>   Holodeck := proc() local Moriarty; Moriarty end proc;
```
$$Holodeck := \mathbf{proc}() \, \mathbf{local} \, Moriarty; \, Moriarty \, \mathbf{end \, proc}$$
```
>   escaped[1] := Holodeck();
```
$$escaped_1 := Moriarty$$

Moriarty has escaped from the Holodeck!
```
>   Moriarty - escaped[1] ;
```
$$Moriarty - Moriarty$$
```
>   escaped[2] := Holodeck();
```
$$escaped_2 := Moriarty$$
```
>   escaped[1] - escaped[2];
```
$$Moriarty - Moriarty$$

He is *different* from any `Moriarty` that the user can type in, or indeed from any other such escaped locals. Different instances of `Moriarty` are different (though they look the same). But nonetheless he can be touched:

```
>  cat(``,escaped[1])-cat(``,escaped[2]);
```
$$0$$

That last trick is due to Dominik Gruntz. We concatenate the NULL string to the local variable name, and the result (a global name) is independent of where the variable came from: Local or global, it doesn't matter.

3.4.2 Global Variables

Global variables are available to anything that references their name. If a local variable has the same name as a global variable, the local variable "masks" (or "shadows") the global one and that global variable is unavailable in that procedure, unless an explicit reference to the "outer" scope is made by using the prefix :-, as follows. The construct :-x refers to the x in the global scope, even if you have a local variable named x. Use of global variables should be restricted as much as possible, since they can conflict with other users' names for their variables (or indeed your own). If you must use a global variable, make its name longer than necessary, to help avoid conflicts. In procedures, global variables should be explicitly declared global. Otherwise, they will be automatically declared as local if they occur on the left-hand side of an assignment or in a loop, and this may not be what was intended. These are Maple's implicit scoping rules.

Most system global variables start with the character _, e.g., _NCRule. Do not use variable names starting with this character, except for "environment" variables; see below.

It is often difficult to make sure you have declared all your variables as local, and unexpected globals can be a time bomb for your routines: They may work fine for months, and suddenly fail if you assign a global variable with the same name as the inadvertent global in your routine. The program utility mint, called outside of Maple,[1] helps to deal with this. It will check to see if there are any global variables in your procedure, and check to see that all locals are used. It checks for other Maple syntax errors as well, which can be much quicker than running Maple to find the syntax errors. So, *use* mint *to check your programs.*

3.4.3 Environment Variables

Environment variables are, roughly speaking, global variables that are automatically reset on exit from a procedure. There are several built-in environment variables in Maple and facilities for adding your own. The built-in environment variables are Digits, Normalizer, Testzero, mod, printlevel, and the percent (last result) commands %, %%, and %%%. Finally, any variable beginning with _Env is an environment variable, which enables you to define your own. See ?environment for more details.

---

[1]The internal routine maplemint has some of the same features.

```
>   restart;
>   fu := proc( x )
>       bah( x );
>       _EnvMyJunk := 3;
>       bah( x );
>   end proc:
>   bah := proc( y )
>       if type(_EnvMyJunk,posint) and _EnvMyJunk=5 then
>           WARNING( "It's 5, I tell you!" )
>       elif type(_EnvMyJunk,posint) and _EnvMyJunk=3 then
>           WARNING( "Look out, it's 3!" )
>       else
>           WARNING( "Everything's cool, it's not 3." )
>       end if;
>   end proc:
>   _EnvMyJunk := 5;
```

$$\_EnvMyJunk := 5$$

```
>   fu( throgmorton );
```

Warning, It's 5, I tell you!
Warning, Look out, it's 3!

```
>   _EnvMyJunk;
```

$$5$$

In the above, we see that _Envmyjunk is reset *automatically* on exit from fu.

The `series` command makes use of the `Testzero` environment variable, and that means that you can select the normalizer to use on computation of series coefficients. This allows correct computation of series in special circumstances, when the generic zero-recognition tools are not strong enough to recognize division by zero.

The following example illustrates the meaning and use of this variable. It also gives our first illustration of the difficulties associated with `option remember`, which will be discussed more fully in Section 3.5.

The *residue* of a function $f(z)$ at a point $z = a$ is the coefficient of $1/(z - a)$ in the Laurent series expansion of f at $z = a$. It is used in the computation of contour integrals, among other things.

```
>   restart;
>   p := x^3 + x + 1;
```

$$p := x^3 + x + 1$$

```
>   alias( alpha=RootOf(p,x) );
```

$$\alpha$$

```
>   residue( 1/p, x=alpha );
```

$$0$$

This is incorrect, since α is a root of p. But residue calls series which uses normal which does not recognize that $p(\alpha) = 0$ (because normal is not strong enough).

```
>    series( 1/p, x=alpha, 2 );
```

$$\frac{1}{\alpha + 1 + \alpha^3} - \frac{1 + 3\alpha^2}{(\alpha + 1 + \alpha^3)^2}(x - \alpha) + O((x - \alpha)^2)$$

That answer is incorrect.

```
>    Testzero := x -> evalb(Normalizer(x)=0);
```

$$Testzero := x \rightarrow \text{evalb}(\text{Normalizer}(x) = 0)$$

```
>    Normalizer := x -> normal(simplify(x));
```

$$Normalizer := x \rightarrow \text{normal}(\text{simplify}(x))$$

```
>    series( 1/p, x=alpha, 2 );
```

$$\frac{1}{\alpha + 1 + \alpha^3} - \frac{1 + 3\alpha^2}{(\alpha + 1 + \alpha^3)^2}(x - \alpha) + O((x - \alpha)^2)$$

It's still wrong. That incorrect result was remembered from before. We must use forget to tell series to throw away that old result.

```
>    forget( series );
>    series( 1/p, x=alpha, 2 );
```

$$\frac{1}{1 + 3\alpha^2}(x - \alpha)^{-1} + O\left((x - \alpha)^0\right)$$

We see now that forget has wiped out the remember table for series.

```
>    residue( 1/p, x=alpha );
```

$$0$$

We need to call forget again, this time on residue.

```
>    forget( residue );
>    residue( 1/p, x=alpha );
```

$$\frac{1}{1 + 3\alpha^2}$$

That's better.

Exercises

1. Compute the residues of $q = (1 + z)/(1 + z + z^2 + z^3)$.

2. Compute the series expansion of q about $z_0 = \exp(i\pi/4)$.

3.4.4 Nested Lexical Scopes

One of the most welcome improvements to the Maple programming language of the past five years was the introduction of nested lexical scopes. Put simply, what this means is that a variable used in a procedure will be expected at first to be in the local scope; that is, a variable local to that procedure. If it is not found in that scope, then it will be expected to be in the next "outer" scope; if not there, then the next outermost one, and so on all the way out until it is found to be global.

The first example of the use of nested lexical scopes in this book was in the program FourierSineSeries in Figure 1.6. The operator that is returned as an answer contains a reference to the variable c, which is local to FourierSineSeries (and therefore uniquely named). Therefore, the operator that is returned uses nested lexical scopes to hide the definition of c from other programs. This means that we can call FourierSineSeries twice on different problems, without the second call destroying the results of the first. This is far superior to the version of FourierSineSeries that was given in the first edition of this book.

The next example in this book was the program veil, in Figure 2.1. Here, the hidden (local) auxiliary function refers to the nested variables lastUsed and computationSequence and the input symbol C. This is the only way to parameterize a module, by building it inside a procedure so that the parameters of the module are inherited by nested lexical scopes.

The next example of the use of nested lexical scopes is the module for finite difference operators in Section 3.2.1. There, for example, the operator E would take as input an operator f, and return an operator $x \rightarrow f(x+h)$, where the x is local to the returned operator, but the h was nested from the procedure that built the module that exported the operator E (two levels out). The exported operator S has a variable k in it that is local to the module (one level out).

The program parsolve in Figure 3.6 also uses nested lexical scopes; otherwise, a cumbersome circumlocution would have to be used (as it was in the first edition of this book).

One problem with nested lexical scopes is that you cannot use them with evalhf. This is why the procedure in Figure 3.8 uses codegen[makeproc] to build a procedure out of one-level evaluated elements.

3.5 Recursion and option remember

> Drink is the curse of the working classes.
>
> > —Anonymous
>
> Work is the curse of the drinking classes.
>
> > —Oscar Wilde
>
> Recursing is the work of the thinking classes.
>
> > —David Jeffrey

Recursion allows the programmer to be lazy. To be fair, it does allow compact programs, and programmer time is important, too. However, although recursive programs can be efficient if the programmer is careful, if the programmer is naive, then recursion can lead to extremely wasteful usage of computer resources (to the point where the desired task cannot be performed). The classical example of this is the recursive calculation of the Fibonacci numbers; a discussion of that example can be found in [44]. Here we look at a similar example, arising from a recursive formulation of the determinants of a class of tridiagonal matrices.

```
>   restart;

>   A := n -> Matrix(n,n,
>           (i,j)->'if'(abs(i-j)=1,1,'if'(i=j,i,0)));
```

$$A := n \rightarrow \text{Matrix}(n, n, (i, j) \rightarrow \text{'if'}(|i - j| = 1, 1, \text{'if'}(i = j, i, 0)))$$

```
>   A(5);
```

$$\begin{bmatrix} 1 & 1 & 0 & 0 & 0 \\ 1 & 2 & 1 & 0 & 0 \\ 0 & 1 & 3 & 1 & 0 \\ 0 & 0 & 1 & 4 & 1 \\ 0 & 0 & 0 & 1 & 5 \end{bmatrix}$$

```
>   use LinearAlgebra in
>           seq( Determinant(A(n)), n=1..8 )
>   end use;
```

$$1, 1, 2, 7, 33, 191, 1304, 10241$$

By doing a Laplace expansion about the final row, we discover the following recurrence relation for the determinants of the n-by-n matrices A_n:

$$\det A_n = n \det A_{n-1} - \det A_{n-2} . \tag{3.1}$$

We can implement this in a naive recursive way as follows:

```
>   dt_naive := proc(n)
>       description "Example recursive program.";
>       if n <=2 then
>           1
>       else
>           n*dt_naive(n-1)-dt_naive(n-2)
>       end if
>   end proc;:

>   seq(dt_naive(n),n=1..8);
```

$$1, 1, 2, 7, 33, 191, 1304, 10241$$

We can use the debugger (see Section 3.8) to trace the execution of this routine, as follows.

```
>   showstat(dt_naive);

dt_naive := proc(n)
   1    if n <= 2 then
   2       1
         else
   3       n*dt_naive(n-1)-dt_naive(n-2)
         end if
end proc
```

It is interesting that the description statement is not even shown, but that other unnumbered statements are present.

```
>   stopat(dt_naive,2);
```

$$[dt\_naive]$$

```
>   dt_naive(5);

dt_naive:
   2*    1


>   where;

TopLevel: dt_naive(5)
[5]
dt_naive: n*dt_naive(n-1)-dt_naive(n-2)
[4]
dt_naive: n*dt_naive(n-1)-dt_naive(n-2)
[3]
dt_naive: n*dt_naive(n-1)-dt_naive(n-2)
[2]
dt_naive:
   2*    1
```

The integers in brackets are the successive inputs to dt_naive.

```
>   cont

dt_naive:
   2*    1


>   where

TopLevel: dt_naive(5)
[5]
dt_naive: n*dt_naive(n-1)-dt_naive(n-2)
[4]
dt_naive: n*dt_naive(n-1)-dt_naive(n-2)
[3]
dt_naive: n*dt_naive(n-1)-dt_naive(n-2)
[1]
dt_naive:
   2*    1
```

```
>   cont

dt_naive:
    2*    1

>   where

TopLevel: dt_naive(5)
[5]
dt_naive: n*dt_naive(n-1)-dt_naive(n-2)
[4]
dt_naive: n*dt_naive(n-1)-dt_naive(n-2)
[2]
dt_naive:
    2*    1

>   cont

dt_naive:
    2*    1

>   where

TopLevel: dt_naive(5)
[5]
dt_naive: n*dt_naive(n-1)-dt_naive(n-2)
[3]
dt_naive: n*dt_naive(n-1)-dt_naive(n-2)
[2]
dt_naive:
    2*    1

>   cont

dt_naive:
    2*    1

>   where

TopLevel: dt_naive(5)
[5]
dt_naive: n*dt_naive(n-1)-dt_naive(n-2)
[3]
dt_naive: n*dt_naive(n-1)-dt_naive(n-2)
[1]
dt_naive:
    2*    1

>   cont
```

33

It is clear that three of those calls were unnecessary repeats of identical earlier calls. If we specify option remember, then we tell Maple to automatically place the values computed by the procedure into its "remember table," and next time not to recompute but simply return what it had computed before. If we don't do this here, then we incur a cost for computing det A_n that grows *exponentially* with n.

```
>   N := 27;
```

$$N := 27$$

```
>   times := array(1..N);
```

$$times := \text{array}(1..27, [])$$

```
>   for i to N do
>      st := time():
>      dt_naive(i);
>      times[i] := time()-st;
>   end do:
>   times[N];
```

$$9.351$$

```
>   times[10];
```

$$0.$$

```
>   times[15];
```

$$.025$$

```
>   times[12];
```

$$.005$$

```
>   times[11];
```

$$.005$$

```
>   dataplot := plots[logplot]([seq([i,(times[i])],
>                  i=11..N)],
>         style=POINT, symbol=BOX, colour=BLACK,
>         axes=BOXED, labels=["n","cputime"] ):
>   plots[display](dataplot);
```

That plot is not shown here. Instead, we fit a straight line to the data, and display the line and data together. We use the stats package to find a line of good fit.

```
>   with(stats):
>   Xvalues := [seq(i,i=11..N)];
```

$$Xvalues := [11, 12, 13, 14, 15, 16, 17, 18, 19,$$
$$20, 21, 22, 23, 24, 25, 26, 27]$$

```
>   Yvalues := [seq(log(times[i]),i=11..N)];
```

$$Yvalues := [-5.298317367, -5.298317367, -4.605170186,$$
$$-4.199705078, -3.688879454, -3.218875825, -2.733368009,$$
$$-1.864330162, -1.766091722, -1.108662625, -.7133498879,$$
$$-.1996711951, .3111544286, .7747271676, 1.268355063,$$
$$1.754403683, 2.235483289]$$

```
>    fit[leastsquare[[n,logT],logT=a*n+b,{a,b}]](
>          [Xvalues,Yvalues] );
```

$$logT = .4892309278\,n - 10.96307088$$

```
>    T := unapply( exp( rhs(%) ), n);
```

$$T := n \to e^{(.4892309278\,n - 10.96307088)}$$

```
>    theoryplot := plots[logplot](T,11..N):
>    plots[display]({dataplot,theoryplot});
```

See Figure 3.4.

The time for dt_naive is essentially proportional to the value of the nth Fibonacci number, which grows exponentially with n. The Maple time() function returns the time in seconds from some reference time.

To predict how long this would take to compute det A_{100}, we use the straightline fit to the logarithmic data from above:

```
>    T( 100 )/(3.156*10^7);
```

$$.9698198066\,10^9$$

This prediction gives roughly 1 billion years, on this machine. The number of seconds in a year is roughly $\pi \cdot 10^7$, which is easy to remember; but here is a more accurate value:

```
>    365.25*24*3600;
```

$$.3155760000\,10^8$$

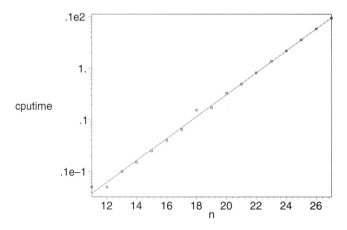

Figure 3.4: Exponential cost of a naive recursive program

Now, how about if we factor in Moore's law? My machine is a year old, and I could lower that cpu time by a factor of about 10 by running the program on the fastest machine available to me; thereafter, computing speeds double every 18 months (roughly), and so after 29 more years of development, Moore's law predicts that we will have computers capable of computing det A_{100} in about a year's worth of computing, by this naive method. Any bets that this will actually happen?

But there is a better alternative. We now examine how option remember turns this exponential growth in computing cost into polynomial (indeed linear) growth.

```
>   dt_ok := proc(n)
>      option remember;
>      description "More efficient recursive program"
>          "with option remember.";
>      if n <=2 then
>         1
>      else
>         n*dt_ok(n-1)-dt_ok(n-2)
>      end if
>   end proc:
```

Note that there is no semicolon between the proc(n) and the option remember.

```
>   showstat(dt_ok);
```

```
dt_ok := proc(n)
   1   if n <= 2 then
   2     1
       else
   3     n*dt_ok(n-1)-dt_ok(n-2)
       end if
end proc
```

The debugger doesn't print the option remember statement, either.

```
>   stopat(dt_ok,2);
```

$$[dt\_naive,\ dt\_ok]$$

```
>   dt_ok(5);
```

```
dt_ok:
   2*    1
```

```
>   where

TopLevel: dt_ok(5)
[5]
dt_ok: n*dt_ok(n-1)-dt_ok(n-2)
[4]
dt_ok: n*dt_ok(n-1)-dt_ok(n-2)
[3]
dt_ok: n*dt_ok(n-1)-dt_ok(n-2)
[2]
dt_ok:
    2*    1

>   cont

dt_ok:
    2*    1

>   where

TopLevel: dt_ok(5)
[5]
dt_ok: n*dt_ok(n-1)-dt_ok(n-2)
[4]
dt_ok: n*dt_ok(n-1)-dt_ok(n-2)
[3]
dt_ok: n*dt_ok(n-1)-dt_ok(n-2)
[1]
dt_ok:
    2*    1

>   cont
```
                                33

Obviously, this takes many fewer calls.

```
>   seq( dt_ok(n), n=1..8 );
```
                1, 1, 2, 7, 33, 191, 1304, 10241

In theory, the computing time needed by this program grows linearly with the input. In fact, the times taken by this program are really too small to register on the clock, for reasonable values of n. Trying to measure the times taken essentially just gives the time needed for Maple to do garbage collections. For example, the time to compute det A_{100} by this method on this machine is reported as 0.003 seconds. This beats a billion years, or even thirty.

option remember is a good thing, occasionally, and a bad thing if overused, because memory requirements are often much higher in a procedure that uses option remember than in a correctly written procedure that doesn't. Compare fib of [44] to the built-in combinat[fibonacci] procedure, which is *much* more efficient, and does *not* use option remember.

As a further note on recursion, there is a built-in (system dependent) limit on the stack, which limits the depth to which Maple procedures can call themselves.

For dt_ok on my machine, it is 755 levels. Of course, because it has option remember, if you want to compute det A_n for $n > 755$, you can do so by dividing $n = 755q + r$ and then issuing q invocations of dt_ok(i*755), for $i = 1, 2, \ldots$, q, gradually building up the remember table, and then finally dt_ok(n). On my machine it takes 0.2 seconds to compute det A_{755} using dt_ok. The prediction of the time dt_naive would take is 10^{144} years, or 500 years of continual Moore's law improvement.

Warning: option remember *can cause bugs.* option remember does *not* take into account the values of mutable variables such as tables, Matrices, Vectors, global or environment variables (except in evalf, which takes special care with Digits). So if your proc uses global or environment variables, *don't* use option remember. This is why we had to use forget for series when we changed the environment variable Testzero earlier.

To repeat, option remember and tables, Arrays, Matrices, or Vectors do not mix. Assignment of table entries does not change the name of the table, hence option remember won't notice if any table entries have been changed.

```
> restart;
> f := proc( t::table )
>    option remember;
>    description "Procedure showing that option remember"
>       " and tables don't mix.";
>    if t[1]=3 then
>       return "The entry is three.";
>    else
>       return "The entry", t[1], " is not three.";
>    end if;
> end proc:

> T := table():

> T[1] := 17;
```
$$T_1 := 17$$

```
> f( T );
```
$$\text{"The entry", 17, " is not three."}$$

```
> T[1] := 3;
```
$$T_1 := 3$$

```
> f( T );
```
$$\text{"The entry", 17, " is not three."}$$

```
> T[1];
```
$$3$$

The procedure f has remembered the previous result, even though it is not what we wanted. *Don't mix tables, arrays, or vectors with* option remember.

Programming notes.

1. We could have used `sprintf` in the return statement to embed the entry `t[1]` into the output string, thus avoiding the ugly commas.

2. If we really want to mix globals, environment variables, or tables with a procedure that has `option remember`, then we can use the following trick. Write an auxiliary procedure that accepts all of the relevant quantities (e.g., `Digits`, `t[1]`, `_EnvX`) as parameters, e.g., `Aux( x, digits, t1, X )`; and give that auxiliary procedure `option remember`, not the outer wrapper that takes the environment variables and calls Aux with them. This is what was done in the module LEM in Figure 2.1.

Final advice on `option remember`: Use it if you want (it can save you time), but be careful. If you can spare the effort, see whether you can write your procedures without it. Also, be aware of `option system`, which modifies `option remember` so that that garbage collection clears the remember table frequently.

3.6   Variable Number or Type of Arguments

The number of actual arguments to Maple `procs` can vary at run time. Some functions must have a minimum number of arguments; for example,

```
>   restart;
>   f := proc( x, y )
>      if x < 1 then
>         1
>      elif y < 3 then
>         2
>      else
>         3
>      end if
>   end proc:
>   f( 0 );
```

$$1$$

```
>   f( 1, 2 );
```

$$2$$

```
>   f( 1 );
```

Error, (in f) f uses a 2nd argument, y, which is missing

If this routine is called with no arguments, by `f()`, then an error message will ensue. If it is called with only one argument, one of two things can happen, depending on the value of the argument, as we saw.

For times when you want to deal with the case where there are *more* arguments than specified in the procedure header, you can use args and nargs. At runtime, the variable nargs contains the number of actual arguments to the procedure (which may vary from invocation to invocation), and the expression sequence args contains the actual arguments: The *i*th argument can be referenced as args[i]. See also the next section for more examples of the use of nargs and args.

Structured types allow type-checking for more than one type of input argument. The following procedure will accept arguments of type name or an equation where the left-hand side is a name and the right-hand side is a range with constant endpoints. This is useful for setting up default cases, for example.

```
>   restart;
```

We are also going to use the typematch utility to pick out the parameters. The following routine supplies a default range of integration.

```
>   my_int := proc( f, x::{name,name=anything..anything} )
>       local a, b, s, t;
>       description "Definite integration, "
>           "by default on 0..1.";
>       if typematch(x,t::name=a::anything..b::anything) then
>           int( f, t=a..b )
>       else
>           int( f, x=0..1 )
>       end if
>   end proc:
```

If the match of the type succeeds, the variables *t*, *a*, and *b* are automatically bound to the correct matching elements, and can be used in the int command that follows.

```
>   my_int( sin(x), x );
```
$$-\cos(1) + 1$$

```
>   my_int( sin(x), x=0..Pi );
```
$$2$$

```
>   my_int( cos(q), q=-2..2 );
```
$$2\sin(2)$$

```
>   my_int( exp(w), 0..1=w );
```

```
Error, invalid input: my_int expects its 2nd argument x to be of
type {name, name = anything .. anything}, but received 0 .. 1 = w
```

If you create lots of Maple programs, you should create help files for each of them. This can now be done easily with the makehelp procedure. See ?makehelp for details, and Section 3.9.1 for an example. Another thing that is useful if you are creating many programs is the Maple Archiver or march. This helps you to manage archives (repositories) of .m files. See ?march and ?repository for details.

3.7 Returning More Than One Result

Maple is set up to return only one result from a procedure. Mind you, that result can be an array, table, list, set, or expression sequence, so the consequences of this restriction are not severe. Now that expression sequences can appear on the left-hand side of an assignment,[2] there really is no need to return values through the parameter list, FORTRAN-style. However, this is possible, and the following example shows how. Pay particular attention to the use here of right, or forward, quotation marks, which prevent premature evaluation of the names to be assigned.

```
>   restart;
```

The procedure mynormal below returns an expression sequence as an answer, the elements of which are supposed to be the numerator and denominator of a quotient. If asked, this routine also returns the gcd of the two quantities.

```
>   mynormal := proc( p::polynom, q::polynom, x::name, GCD )
>      local g, pc, qc;
>      description "Example of passing results"
>            " back through parameters.";
>      g := gcd(p, q);
>      if nargs > 3 and type( GCD, name ) then
>         GCD := g;
>      end if:
>      rem( p, g, x, pc );
>      rem( q, g, x, qc );
>      (pc, qc)
>   end:
```

This is a "recursive" example, since the subroutine rem (and its complementary routine quo) use the trick we are trying to explain here. See ?rem for details.

```
>   (num, den) := mynormal( 1+s, 1+2*s+s^2, s );
```
$$num, den := 1, 1+s$$

```
>   (num, den) := mynormal( 1+s, 1+2*s+s^2, s, 'g' );
```
$$num, den := 1, 1+s$$

```
>   g;
```
$$1+s$$

```
>   P := expand( (x+1)^3*(x+2)^2*(x+3) );
```
$$P := x^6 + 10\,x^5 + 40\,x^4 + 82\,x^3 + 91\,x^2 + 52\,x + 12$$

---

[2]It is considered good style to wrap such expressions in parentheses: $(A, B) := (1, 2)$.

```
>   Q := diff( P, x );
```

$$Q := 6\,x^5 + 50\,x^4 + 160\,x^3 + 246\,x^2 + 182\,x + 52$$

```
>   (num, den) := mynormal( P, Q, x, g );
```

$$num,\ den := x^3 + 6\,x^2 + 11\,x + 6,\ 6\,x^2 + 26\,x + 26$$

```
>   g;
```

$$1 + s$$

```
>   (num, den) := mynormal( P, Q, x, 'g' );
```

$$num,\ den := x^3 + 6\,x^2 + 11\,x + 6,\ 6\,x^2 + 26\,x + 26$$

```
>   g;
```

$$x^3 + 4\,x^2 + 5\,x + 2$$

The previous time, even though there were four arguments present, the fourth argument was not of type name, and so no assignment was attempted, which left *g* alone. If the assignment had been attempted, we would have received an error message. Perhaps it is desirable that an error message be generated because g might inherit a previous value, erroneous in the present context. See ?error. That help page describes how to use the error statement, and in particular how to pass a parameter into the message string using %1.

```
>   mynormal2 := proc(p::polynom, q::polynom, x::name, GCD)
>      local g, pc, qc;
>      description "Improved pass-back example, with"
>                  "user-defined error message.";
>      g := gcd(p, q);
>      if nargs > 3 then
>         try
>            GCD := g;
>         catch :
>            error "I'm very sorry, "
>                  "I cannot assign to \"%1\"", GCD ;
>         end try;
>      end if:
>      rem( p, g, x, pc );
>      rem( q, g, x, qc );
>      (pc, qc)
>   end:
```

```
>   G := "Something we need.";
```

$$G := \text{``Something we need.''}$$

```
>   mynormal2( P, Q, x, G );
```

```
Error, (in mynormal2) I'm very sorry, I cannot assign to
"Something we need."
```

```
>   G;
```

$$\text{``Something we need.''}$$

```
>   mynormal2( P, Q, x, 'G' );
```
$$x^3 + 6\,x^2 + 11\,x + 6,\ 6\,x^2 + 26\,x + 26$$
```
>   G;
```
$$x^3 + 4\,x^2 + 5\,x + 2$$

A useful strategy to deal with code like the original `mynormal` (some library routines use this method of passing values; e.g., `rem` and `quo`) is to pass unevaluated names. That is, use quotation marks whenever there is a chance that the names may have values. In Figure 2.11 the keyword `integer` was quoted this way. Notice that in the calls to `rem` in `mynormal` and `mynormal2` quotation marks around `pc` and `qc` are not used: This is because they are unnecessary since the local variables obviously have no other values and on each invocation get created anew.

Even so, if this code were later to be modified, it is possible that values could be assigned to these local variables; in that case, an error would occur. We saw above an example of quoting names in the input. This "unevaluation" said to `mynormal` to use the *name* g and not its value. Of course, this works for `mynormal2`, also.

An alternative to using quotation marks, that is more convenient for the user and programmer, but has some risk of accidentally overwriting data without warning, is to declare the type of the return parameter to be `evaln`. See [44] for details.

3.8   Debugging Maple Programs

The debugging facilities of Maple include `tracelast`, which we have already seen; `mint`, an external program that does syntax-checking on Maple programs; `trace`, which traces execution of a program; `printlevel`, which allows you to watch the execution of a program if you set it to a high enough integer; and (most important and most recent) the Maple debugger, which allows you to set watchpoints and breakpoints in a program and to single-step your way through it.

As an example, in Figure 3.5 we find a putative program to compute the Jacobian matrix of a list of input functions. It contains one deliberate error.

```
>   restart;
>   read "D:/books/ess/programs/jacobian.mpl";
>   f := [ sin(x-y), cos(x) ];
```
$$f := [\sin(x - y),\ \cos(x)]$$
```
>   J := Jacobian( f, [x,y] );
```
$$J := \left[\begin{array}{cc} \cos(x - y) & \cos(x - y) \\ 0 & 0 \end{array} \right]$$

Now, it is apparent that that is wrong. So we step through the code to see what is happening.

```
Jacobian := proc( fnlist :: list, varlist :: list(name) )
   local J, i, j, m, n, tmp, inds;
   description "Sparse Jacobian Matrix of list of functions.";
   m := nops( fnlist );
   n := nops( varlist );
   J := Matrix( m, n, storage=sparse );
   for i to m do
      inds := indets( fnlist[i] );
      for j to n do
        if member( varlist[j], inds ) then
           # The deliberate bug is that varlist[i] should be [j]
           tmp := Normalizer( diff( fnlist[i], varlist[i] ) );
           if not Testzero( tmp )  then
              J[i,j] := tmp;
           end if;
        end if;
      end do;
   end do;
   J
end proc:
```

Figure 3.5: A deliberate bug in a program for Jacobians

```
>   showstat( Jacobian );


Jacobian := proc(fnlist::list, varlist::list(name))
local J, i, j, m, n, tmp, inds;
   1*  m := nops(fnlist);
   2   n := nops(varlist);
   3   J := Matrix(m,n,storage = sparse);
   4   for i to m do
   5     inds := indets(fnlist[i]);
   6     for j to n do
   7       if member(varlist[j],inds) then
   8         tmp := Normalizer(diff(fnlist[i],varlist[i]));
   9         if not Testzero(tmp) then
  10           J[i,j] := tmp
             end if
           end if
         end do
       end do;
  11   J
end proc


>   J := Jacobian( f, [x,y] );

Jacobian:
   1*  m := nops(fnlist);
```

```
>   next
```

```
2
Jacobian:
   2    n := nops(varlist);
```

```
>   next
```

```
2
Jacobian:
   3    J := Matrix(m,n,storage = sparse);
```

```
>   next
```

```
Matrix(2, 2, [[0,0],[0,0]], datatype = anything, storage = sparse,
order = Fortran_order, shape = [])
Jacobian:
   4    for i to m do
              . . .
         end do;
```

```
>   into
```

```
Matrix(2, 2, [[0,0],[0,0]], datatype = anything, storage = sparse,
order = Fortran_order, shape = [])
Jacobian:
   5      inds := indets(fnlist[i]);
```

```
>   next
```

```
{x, y, sin(x-y)}
```

At this point, a potential bug that I didn't know about has surfaced: I really should have restricted this list to names only! I should replace the line

```
inds := indets(fnlist[i]);
```

with

```
inds := select( type, indets(fnlist[i]), name );
```

Now on to find the deliberate bug . . .

```
Jacobian:
   6      for j to n do
              . . .
         end do
```

```
>  list

Jacobian := proc(fnlist::list, varlist::list(name))
local J, i, j, m, n, tmp, inds;
   1*  m := nops(fnlist);
   2   n := nops(varlist);
   3   J := Matrix(m,n,storage = sparse);
   4   for i to m do
   5      inds := indets(fnlist[i]);
   6 !    for j to n do
   7         if member(varlist[j],inds) then
              ...
            end if
            ...
         end do
         ...
      end do;
      ...
end proc
```

The exclamation mark (!) shows where we are in the procedure.

```
>  into

{x, y, sin(x-y)}
Jacobian:
   7         if member(varlist[j],inds) then
              ...
            end if


>  fnlist[i];

sin(x-y)
Jacobian:
   7         if member(varlist[j],inds) then
              ...
            end if


>  varlist[i];

x
Jacobian:
   7         if member(varlist[j],inds) then
              ...
            end if


>  varlist[j];

x
Jacobian:
   7         if member(varlist[j],inds) then
              ...
            end if
```

```
>  into

{x, y, sin(x-y)}
Jacobian:
    8           tmp := Normalizer(diff(fnlist[i],varlist[i]));
```

Now we look at what goes into that expression, because we suspect that this
is the error:

```
>  next

cos(x-y)
Jacobian:
    9           if not Testzero(tmp) then
                    ...
                end if


>  next

cos(x-y)
Jacobian:
    7           if member(varlist[j],inds) then
                    ...
                end if


>  outfrom

cos(x-y)
Jacobian:
    5       inds := indets(fnlist[i]);


>  list


Jacobian := proc(fnlist::list, varlist::list(name))
local J, i, j, m, n, tmp, inds;
    1*  m := nops(fnlist);
    2   n := nops(varlist);
    3   J := Matrix(m,n,storage = sparse);
    4   for i to m do
    5 !   inds := indets(fnlist[i]);
    6     for j to n do
              ...
            end do
            ...
          end do;
          ...
end proc


>  i;

2
Jacobian:
    5       inds := indets(fnlist[i]);
```

```
>  j;

3
Jacobian:
   5     inds := indets(fnlist[i]);


>  into

{x, cos(x)}
Jacobian:
   6     for j to n do
             ...
         end do


>  into

{x, cos(x)}
Jacobian:
   7        if member(varlist[j],inds) then
               ...
           end if


>  into

{x, cos(x)}
Jacobian:
   8           tmp := Normalizer(diff(fnlist[i],varlist[i]));


>  i;

2
Jacobian:
   8           tmp := Normalizer(diff(fnlist[i],varlist[i]));


>  j;

1
Jacobian:
   8           tmp := Normalizer(diff(fnlist[i],varlist[i]));
```

It's all just as we asked; but we can (if we think a little) realize at this point that we are differentiating with respect to the wrong variable. So we should change `varlist[i]` to `varlist[j]`.

```
>  into

0
Jacobian:
   9           if not Testzero(tmp) then
                 ...
               end if
```

```
>  outfrom

0
Jacobian:
   7          if member(varlist[j],inds) then
                   ...
              end if


>  list


Jacobian := proc(fnlist::list, varlist::list(name))
local J, i, j, m, n, tmp, inds;
       ...
   2   n := nops(varlist);
   3   J := Matrix(m,n,storage = sparse);
   4   for i to m do
   5     inds := indets(fnlist[i]);
   6     for j to n do
   7 !     if member(varlist[j],inds) then
   8         tmp := Normalizer(diff(fnlist[i],varlist[i]));
              ...
            end if
            ...
          end do
          ...
        end do;
        ...
end proc


>  outfrom

0
Jacobian:
  11    J


>  list


Jacobian := proc(fnlist::list, varlist::list(name))
local J, i, j, m, n, tmp, inds;
       ...
   6     for j to n do
   7       if member(varlist[j],inds) then
   8         tmp := Normalizer(diff(fnlist[i],varlist[i]));
   9         if not Testzero(tmp) then
  10           J[i,j] := tmp
              end if
            end if
          end do
        end do;
  11 ! J
end proc
```

```
>   cont
```

$$J := \begin{bmatrix} \cos(x - y) & \cos(x - y) \\ 0 & 0 \end{bmatrix}$$

See `?debugger` for details on these commands.

Exercises

1. Write a procedure that prints out (in a nice format) the amount of CPU time Maple has used so far, together with a report of how much memory it has used. See `?status`.

2. Write a procedure `poleplot` to compute and plot all the poles of a rational function.

3. Write a procedure `zeroplot` to compute and plot all the zeros of a rational function.

3.9 Sample Maple Programs

What follows are a few final sample Maple programs that you may use as templates for your own programs and that may give you ideas for other programs. They are not intended to be examples of "programming gems," and I offer them only as working programs that I wrote for actual use.

3.9.1 Parametric Solution of Algebraic Equations

The following procedure uses the trick of substituting $y = tx$ into an algebraic equation $f(x, y) = 0$ to get a parametric solution of the equation. This is useful for plotting purposes or for integration, differentiation, and series. For more on finding parametric solutions of algebraic equations, see [51].

The program is given in Figure 3.6. The help file for it, which was created as a worksheet and included into my version of Maple by using `makehelp`, is in Figure 3.7. The input to this procedure is the equation to be solved, f, represented as an expression in two variables, which are referred to as x and y in the procedure but, of course, one may use any distinct names for the actual arguments. We also input the actual names of the variables as a list or a set of names, and then finally the name of the parameter (represented as t in the procedure).

We use `typematch` to distinguish between the two kinds of possible input, and to provide a default point of expansion.

We then solve the equation with y replaced by $y_0 + t(x - x_0)$ for x. The call to `solve` is wrapped in set brackets, so multiple solutions returned by `solve` will be pared down to only one solution and the NULL solution will be transformed to the empty set. We subtract the solution $x = x_0$ from this solution set because

```
parsolve := proc( f,
    xy::{ [name,name], [name=anything,name=anything] },
    t::name )
  local p, x, y, x0, y0, s;
  description "Parametric solution of f(x,y)=0.";

  if not typematch( xy, [x::name=x0::anything,y::name=y0::anything] ) then
    x  := xy[1];
    x0 := 0;
    y  := xy[2];
    y0 := 0;
  end if;

  # Throw away the point solution, if it is there, because it
  # is not interesting: we are looking for curves, not points.

  p := {solve( eval(f, y=y0+t*(x-x0)), x)} minus {x0};

  seq( {x=Normalizer(xi), y=Normalizer(y0+t*(xi-x0))}, xi=p )

end proc:
```

Figure 3.6: A Maple program to solve $p(x, y) = 0$ parametrically

the solution $x = x_0$, $y = y_0$, if it occurs, is always uninteresting in this context (because it describes a point, not a curve).

The final use of seq to construct an expression sequence of solutions uses nested lexical scopes (to get the right t, x, and y) and the general iterator form of seq that allows us to iterate over a range, set, list, or sequence.

Here are some examples of this procedure's use.

```
>   restart;
```

```
>   read "D:/books/ess/programs/parsolve.mpl";
```

```
>   eq1 := u^2 + v^2 = a^2;
```

$$eq1 := u^2 + v^2 = a^2$$

```
>   sol1 := parsolve( eq1, [u,v], t);
```

$$sol1 := \left\{ v = \frac{t\,a}{\sqrt{1+t^2}},\ u = \frac{a}{\sqrt{1+t^2}} \right\},$$
$$\left\{ v = -\frac{t\,a}{\sqrt{1+t^2}},\ u = -\frac{a}{\sqrt{1+t^2}} \right\}$$

```
>   seq( normal( eval( eq1, sol )), sol={sol1} );
```

$$a^2 = a^2,\ a^2 = a^2$$

parsolve—solve f(x,y)=0 parametrically
Calling Sequence

parsolve(f(x,y), [x=a, y=b], t)

Parameters

f(x,y) — equation to be set equal to zero and solved

[x,y] — list of variables to solve for, or optionally

[x=x0,y=y0] — location of point [x0,y0] to base parameterization on.

t — name of parameter to use in the solution x(t), y(t).

References

G. H. Hardy, *Pure Mathematics,* Cambridge University Press, 1952.

Description

This routine substitutes y = y0 + t(x-x0) into the given function and tries to solve the resulting equation (set equal to zero) for x as a function of t. If successful, this gives a parametric solution of the original equation, in that f(x(t), y0 + t*(x(t)-x0)) = 0.

Known weaknesses: will not find solutions of the form x = x0 + a*t, y = y0+b*t.

Examples

```
>  restart;
>  parsolve( u^2 + v^2 - 1, [u=-1,v=0], s );
```

$$\{u = -\frac{-1+s^2}{s^2+1},\ v = 2\,\frac{s}{s^2+1}\}$$

```
>  tacnode := 2*x^4 - 3*x^2*y + y^4 - 2*y^3 + y^2:
>  tacsol := parsolve( tacnode, [x,y], t );
```

$$tacsol := \{y = \frac{1}{2}\,\frac{t^2\,(3+2\,t^2+\%1)}{2+t^4},\ x = \frac{1}{2}\,\frac{(3+2\,t^2+\%1)\,t}{2+t^4}\},$$

$$\{x = \frac{1}{2}\,\frac{(3+2\,t^2-\%1)\,t}{2+t^4},\ y = \frac{1}{2}\,\frac{t^2\,(3+2\,t^2-\%1)}{2+t^4}\}$$

$$\%1 := \sqrt{1+12\,t^2}$$

```
>  parsolve( x^x - y^y, [x,y], t );
```

$$\{y = t\,e^{\left(\frac{\ln(\frac{1}{t})\,t}{-1+t}\right)},\ x = e^{\left(\frac{\ln(\frac{1}{t})\,t}{-1+t}\right)}\}$$

See Also

solve

Figure 3.7: The help file for `parsolve`

So this procedure allows us to find a pair of parametric representations of a circle of radius a. Knowing that trigonometric functions give us a better representation, we could, if we desired, set $\cos\theta = 1/\sqrt{1+t^2}$ to get a better parameterization from this one.

Alternatively, we can obtain the solution about the point $(-a, 0)$, which is known to produce a rational parameterization (because the curve is of genus zero).

```
>   sol2 := parsolve( eq1, [u=-a,v=0], s );
```

$$sol2 := \left\{ v = 2\,\frac{s\,a}{s^2+1}, \; u = -\frac{a\,(s^2-1)}{s^2+1} \right\}$$

```
>   seq( normal( eval( (lhs-rhs)(eq1), sol ),expanded),
>        sol={sol2} );
```

$$0$$

We can also do circles centred elsewhere:

```
>   eq3 := s^2 + s*t + t^2 = a^2;
```

$$eq3 := s^2 + s\,t + t^2 = a^2$$

```
>   parsolve( eq3, [s=0,t=a], u);
```

$$\left\{ t = -\frac{a\,(-1+u^2)}{1+u+u^2}, \; s = -\frac{a\,(1+2\,u)}{1+u+u^2} \right\}$$

And now the folium of Descartes:

```
>   Folium := parsolve( x^3 - 3*a*x*y + y^3 = 0, [x,y], t);
```

$$Folium := \{x = 3\,\frac{t\,a}{1+t^3}, \; y = 3\,\frac{t^2\,a}{1+t^3}\}$$

If we wish to plot it, we must nondimensionalize. The plot below is a bit rough, and is not printed here. It is left to the exercises to explain why the plot is so rough.

```
>   plot( eval( [x/a, y/a, t=-5..5], Folium ),
>         view=[-2..2,-2..2],
>         colour=black, scaling=CONSTRAINED );
```

Now a generalization of that folium.

```
>   Foley := parsolve( x^5 - 5*x*y^3 + y^5, [x,y], u);
```

$$Foley := \{x = 5\,\frac{u^3}{1+u^5}, \; y = 5\,\frac{u^4}{1+u^5}\}$$

Now we show some limitations of this approach. We attempt to solve a random degree-5 polynomial in x and y.

```
>   _EnvExplicit := false;
```

$$\_EnvExplicit := false$$

```
> f := randpoly([x,y], degree=5, sparse);
```
$$f := 54 - 5\,y + 99\,x^3 - 61\,x^2\,y - 50\,x^3\,y - 12\,x^5$$
```
> simpler := parsolve( f, [x,y], s );
```
$$\begin{aligned} simpler := \{x = &\text{RootOf}(12\,\_Z^5 + 50\,\_Z^4\,s \\ &+ (-99 + 61\,s)\,\_Z^3 + 5\,s\,\_Z - 54), \\ y = s\,&\text{RootOf}(12\,\_Z^5 + 50\,\_Z^4\,s \\ &+ (-99 + 61\,s)\,\_Z^3 + 5\,s\,\_Z - 54)\} \end{aligned}$$

That parameterization is in terms of a degree-5 polynomial (the original one, in fact). Therefore the method has failed to tell us anything we didn't know. Another type of failure is discussed in the exercises.

Exercises

1. By carefully investigating the parameterization of the folium of Descartes above, decide why the Maple plot was so rough.

2. In Chapter 2, you were asked to plot the cissoid of Diocles, whose rectangular equation is

$$y^2 = \frac{x^3}{2a - x}$$

and whose standard parametric equations are $x = 2a\sin^2\theta$, $y = 2a\sin^3\theta/\cos\theta$. Try `parsolve` on this problem, and comment.

3. Use `parsolve` to find the nontrivial parametric solution of $y^x = x^y$. Euler was the first to discover this [38]. Why did `parsolve` fail to find the trivial solution $y = x$? Show that it will always fail to find solutions of the form $x = x_0 + at$, $y = y_0 + bt$. Improve the program so that it does find such solutions.

4. Write to me if you have found better or more general tricks for solving equations parametrically.

3.9.2   Path Following in $p(x,y) = 0$

The following example shows one method of generating, from symbolic input, a program to be used in the numerical solution of differential equations. Specifically, the problem we study is how to follow smooth complex paths $(x(s), y(s))$ satisfying $p(x(s), y(s)) = 0$, for a given bivariate function p, assumed analytic in each variable. This corresponds to finding a numerical parameterization of the solution, in contrast to the previous section where we were looking for a symbolic parameterization. This is how one part of algcurves[plot_real_curve] works, by the way.

To derive the equations, consider differentiating $p(x(s), y(s)) = 0$ with respect to s:

$$p_x \dot{x} + p_y \dot{y} = 0 , \tag{3.2}$$

If we assume that $p_y \neq 0$, then it is true that $\dot{x} = -\alpha \exp(i\theta) p_y(x, y)$ for *some* real α and θ; putting that into equation (3.2), we get

$$\dot{x} = -\alpha e^{i\theta} p_y \tag{3.3}$$

$$\dot{y} = \alpha e^{i\theta} p_x . \tag{3.4}$$

If it happens that $p_y = 0$, but $p_x \neq 0$, we can derive the same equations by assuming (3.4) instead to start. In any event, once derived, we can see that paths satisfying these differential equations for *any* smooth choice of $\alpha(x, y)$ and $\theta(x, y)$, with an initial point on the surface $p(x, y) = 0$, will remain in the surface $p(x(s), y(s)) = 0$.

For simplicity we here take $\theta = 0$ but, in fact, we may choose θ to be an arbitrary (smooth) function of x, y, and s. We can ensure that the parameter s is the arc length along the curve by choosing

$$\alpha = \left(|p_x|^2 + |p_y|^2 \right)^{-1/2}$$

$$= \left(u_{x_r}^2 + u_{x_i}^2 + u_{y_r}^2 + u_{y_i}^2 \right)^{-1/2} \tag{3.5}$$

where $p = u + iv$ and we have used the Cauchy–Riemann equations (p is assumed separately analytic in x and in y) to write equations (3.3–3.4) in a purely real form. This gives the real equations

$$\dot{x}_r = \alpha u_{y_r} ,$$

$$\dot{x}_i = -\alpha u_{y_i} ,$$

$$\dot{y}_r = -\alpha u_{x_r} ,$$

$$\dot{y}_i = \alpha u_{x_i} . \tag{3.6}$$

The program to generate, from the input p, a program for these equations that is suitable for use in dsolve/numeric is shown in Figure 3.8.

Programming Notes

1. We want the generated procedure to run under evalhf. This forbids us from using nested lexical scopes. Therefore, we must construct our procedure using codegen[makeproc], and make sure to evaluate the polynomial derivatives one level to pick up the local variables. This relies on carefully matching the scopes of the variables.

2. `codegen[makeproc]` takes a list of equations and turns them into a straight-line program. To include general programming constructs (e.g. loops) you must use the internal representation (`intrep`) instead.

Examples. Note that the variable `xi` pretty-prints as ξ. This is amusing but harmless.

```
> restart;
> read "D:/books/ess/programs/pathDE.mpl";
> f := pathDE( x^2+y^2-1, x, y );
```

$$f := \mathbf{proc}(N,\, t,\, xy,\, ypvec)$$
$$\mathbf{local}\, xr,\, \xi,\, yr,\, yi,\, \alpha,\, u,\, u\_yi,\, u\_yr,\, u\_xi,\, u\_xr;$$
$$xr := xy_1;$$
$$\xi := xy_2;$$
$$yr := xy_3;$$
$$yi := xy_4;$$
$$u := -1 + xr^2 - \xi^2 + yr^2 - yi^2;$$
$$u\_xr := 2*xr;$$
$$u\_xi := -2*\xi;$$
$$u\_yr := 2*yr;$$
$$u\_yi := -2*yi;$$
$$\alpha := 1/\mathrm{sqrt}(u\_xr^2 + u\_xi^2 + u\_yr^2 + u\_yi^2);$$
$$ypvec_1 := \alpha * u\_yr;$$
$$ypvec_2 := -\alpha * u\_yi;$$
$$ypvec_3 := -\alpha * u\_xr;$$
$$ypvec_4 := \alpha * u\_xi$$

end proc

```
> sol := dsolve( numeric, procedure=f,
>        range=-Pi..Pi, start=0,
>        initial=array([1,0,0,0]),
>        procvars=[xr(s),xi(s),yr(s),yi(s)] );
```

$$sol := \mathbf{proc}(rkf45\_x) \ldots \mathbf{end\ proc}$$

```
> plots[odeplot]( sol, [xr(s),yr(s)],
>        scaling=CONSTRAINED, colour=BLACK );
```

See Figure 3.9.

```
> sol := dsolve( numeric, procedure=f,
>        range=-2*Pi..2*Pi, start=0,
>        initial=array([0,1,sqrt(2),0]),
>        procvars=[xr(s),xi(s),yr(s),yi(s)] );
```

$$sol := \mathbf{proc}(rkf45\_x) \ldots \mathbf{end\ proc}$$

```
# pathDE
#   Create a procedure for computation of arc-length paths satisfying
#   p(x,y) = 0 in the double complex plane.
#
# Input:
#   p    ---  bivariate analytic function
#   x, y ---  the variables
#
# Output:
#   proc --- a procedure f( N, t, y, yp ) to pass to dsolve/numeric
#            for numerical solution of
#            xr' =  alpha * u_yr         xi' = -alpha * u_yi
#            yr' = -alpha * u_xr         yi' =  alpha * u_xi
#
#            where alpha = 1/sqrt( u_xr^2 + u_xi^2 + u_yr^2 + u_yi^2 )
#            ensures that the parameter is arc length.
#
pathDE := proc( p, x::name, y::name )
  local u, u_xr, u_xi, u_yr, u_yi, xr, xi, yr, yi;
  description "Numerical parametric solution of p(x,y)=0.";

  u := evalc( Re( eval(p, [x=xr+I*xi,y=yr+I*yi]) ) );

  u_xr := diff( u, xr );
  u_xi := diff( u, xi );
  u_yr := diff( u, yr );
  u_yi := diff( u, yi );

  codegen[makeproc](
    [
    'xr' = 'xy'[1],
    'xi' = 'xy'[2],
    'yr' = 'xy'[3],
    'yi' = 'xy'[4],
    'u'     = u,
    'u_xr'  = u_xr,
    'u_xi'  = u_xi,
    'u_yr'  = u_yr,
    'u_yi'  = u_yi,
    'alpha' = '1/sqrt( u_xr^2 + u_xi^2 + u_yr^2 + u_yi^2) ',
    'ypvec'[1] = 'alpha*u_yr',
    'ypvec'[2] = '-alpha*u_yi',
    'ypvec'[3] = '-alpha*u_xr',
    'ypvec'[4] = 'alpha*u_xi'
    ],

    parameters= [N, t, xy, ypvec],

    locals = [xr, xi, yr, yi, alpha ]
    )
end proc:
```

Figure 3.8: A Maple program to generate another program

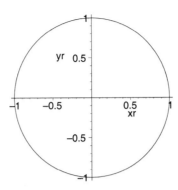

Figure 3.9: Numerical parameterization of the unit circle

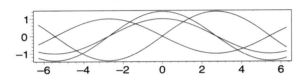

Figure 3.10: A path in the complex circle

```
>    plots[odeplot]( sol,
>             [[s,xr(s)],[s,xi(s)],[s,yr(s)],[s,yi(s)]],
>             scaling=CONSTRAINED, axes=BOXED,
>             labels=["",""], colour=BLACK);
```
See Figure 3.10.
```
>    f2 := pathDE( y^2*exp(-y)-x, x, y )%

>    sol2 := dsolve( numeric, procedure=f2,
>             range=-4..4, start=0,
>             initial=array([0,0,0,0]),
>             procvars=[xr(s),xi(s),yr(s),yi(s)] );
```
$$sol2 := \mathbf{proc}(rkf45\_x) \ldots \mathbf{end\ proc}$$
```
>    plots[odeplot]( sol2, [xr(s),yr(s)],
>             colour=BLACK );
```
See Figure 3.11.
```
>    plots[odeplot]( sol2,
>             [s,abs((xr(s)+I*xi(s)
>                 -(yr(s)+I*yi(s))^2*exp(-(yr(s)+I*yi(s)))))],
>             labels=["",""], style=POINT, colour=BLACK,
>             symbol=CIRCLE, symbolsize=15 );
```

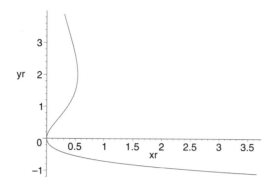

Figure 3.11: Numerical parameterization of a transcendental equation

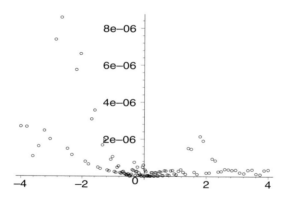

Figure 3.12: The residual error in the computed parameterization

See Figure 3.12.

Exercises

1. Modify the program so that its result computes only real arc-length parameterizations of $p(x, y) = 0$.

2. Draw the solutions to $y^x - x^y = 0$ using pathDE. Compare this to the symbolic parameterization from parsolve.

3.9.3 Large Expression Management, Revisited

In Figure 2.1, we saw a program to help with the management of large expressions. Now that we have seen examples of nested lexical scopes, and option remember, and modules, and modules that are parameterized via nested lexical scopes, it is worthwhile to look over that program again.

The procedure LEM generates a module, with three exports: veil, unveil, and the counter lastUsed. The resulting module is parameterized by the symbol that LEM was called with, and therefore produces labels based on that symbol. The program was written in an object-oriented style, so that the objects exported have state (represented by lastUsed) and behaviour. The internal mechanisms by which it works (a table to store the computation sequence of expressions that are labelled) are hidden from the user.

The counter lastUsed can be modified, by simple assignment (for example if the user wants to use the symbols K_{100} through K_{199}), and so this is not a true object-oriented model; there should be a query function and a setting function to alter the state of the system. I felt that as an example we didn't need any more complexity here.

Here option remember is not used so much for efficiency as to avoid relabelling expressions. It is important in many applications to recognize when two objects are related in a simple manner.

I generated a parameterized warning string to be as informative as possible; moreover, I am still unhappy with the procedure name (LEM) and may change it. Therefore, I parameterized the warning string with a macro that refers to the actual procedure name. Thus, I have only to change the macro later if I decide to change the name.

Finally, we can now talk about the use of the module export syntax. If we bind the exports at the global level by using with on a module that has been generated by LEM, then we will not be able to use two or more sets of labels in the same session. That is why, in the example that I gave of using collect, I generated the module for K and called it VK but did not bind its exports using with; then generating a module for the symbol C and calling it VC allows us to veil or unveil objects with VC:-veil, count how many K constants there are with VK:-lastUsed, and so on.

3.9.4 Fourier Sine Series, Revisited

In Figure 3.13 we find another program to compute the Fourier sine series of a given odd function $f(x)$ on the interval $0 \le x \le 1$. The most important thing that is new about this version, compared to that in Figure 1.6, is that it uses the try and catch statements to try the simple evaluation of the Fourier coefficents, and only if that fails (and the division by zero exception is raised thereby) is the more expensive limit computation used. This exception-handling mechanism is very useful. See ?try, and the discussion in [44, Chap. 7].

```
FourierSineSeries3 := proc( f, x )
  local k, z, F, fop, desc;
  description "More efficient Fourier Sine Series generator.";
  desc := sprintf( "Computes Fourier Sine Series of %a", f );
  fop := unapply(f,x);
  F := 2*int( sin(k*Pi*z)*fop(z), z=0..1 );
  subs(DUMMY=desc,proc(n)
    local j, s, tmp;
    option Copyright;
    description DUMMY;
    s := 0;
    for j to n do
      try
        tmp := eval( F, k=j );
      catch "numeric exception: division by zero":
        tmp := limit( F, k=j );
      finally
        s := s + tmp*sin(j*Pi*x);
      end try;
    end do;
  end proc);
end proc:
```

Figure 3.13: Another Maple program to compute Fourier sine series

Two pieces of "syntactic sugar" of note here are, first, that the `Copyright` option is used to inhibit printing of the (uninformative) procedure body, and, second, that the `description` option is programmatically generated using a formatted string print and substituted into the generated procedure body, so as to provide to the user information about the function whose series is being computed. We cannot use nested lexical scopes to do this, in Maple 7 at least, and so we substitute for a DUMMY global variable. We assume that the procedure in Figure 3.13 has been read into Maple, for the following examples.

>   `macro( FSS3=FourierSineSeries3 );`

$$FSS3$$

>   `f := FSS3( x^2*(1-x^2), x );`

$$f := \mathbf{proc}(n)$$
$$\mathbf{description}$$
"Computes Fourier Sine Series of x^2*(1-x ^2)"
$$\cdots$$
$$\mathbf{end\ proc}$$

Notice that the description field gives useful information, and that the rest of the procedure body is hidden (unless the user raises `verboseproc` to 2).

```
>  f5 := f( 5 );
```

$$f5 := -4\,\frac{(24 - 4\,\pi^2)\sin(\pi\,x)}{\pi^5} - \frac{3\sin(2\,\pi\,x)}{\pi^3}$$

$$-\,\frac{4}{243}\,\frac{(24 - 36\,\pi^2)\sin(3\,\pi\,x)}{\pi^5} - \frac{3}{8}\,\frac{\sin(4\,\pi\,x)}{\pi^3}$$

$$-\,\frac{4}{3125}\,\frac{(24 - 100\,\pi^2)\sin(5\,\pi\,x)}{\pi^5}$$

```
>  fc := FSS3( cos(Pi*x), x );
```

$fc := \mathbf{proc}(n)$
description "Computes Fourier Sine Series of cos(Pi*x)"
$\cdots$
end proc

```
>  fc( 10 );
```

$$\frac{8}{3}\,\frac{\sin(2\,\pi\,x)}{\pi} + \frac{\frac{16}{15}\sin(4\,\pi\,x)}{\pi} + \frac{\frac{24}{35}\sin(6\,\pi\,x)}{\pi} + \frac{\frac{32}{63}\sin(8\,\pi\,x)}{\pi}$$

$$+\,\frac{\frac{40}{99}\sin(10\,\pi\,x)}{\pi}$$

```
>  plot( fc(10)-cos(Pi*x), x=0..1 );
```
See Figure 3.14.

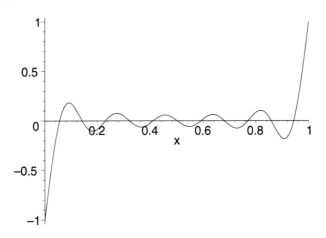

Figure 3.14: The error in 10 terms of the series for $\cos \pi x$

```
>   read "D:/books/ess/programs/fouriersine.mpl";
```

That reads in the routine from Figure 1.6.

```
>   macro( FSS=FourierSineSeries );
```

$$FSS3, FSS$$

```
>   f1 := FSS( x*sin(Pi*x), x );
```

$$f1 := n \to \text{add}(c(k)\sin(k\,\pi\,x),\ k = 1..n)$$

```
>   f1( 10 );
```

```
Error, (in c) numeric exception: division by zero
```

```
>   f3 := FSS3( x*sin(Pi*x), x );
```

$f3 := \textbf{proc}(n)$
$\quad\textbf{description}$ "Computes Fourier Sine Series of x*sin(Pi*x)"
$\qquad\cdots$
$\quad\textbf{end proc}$

```
>   f3( 10 );
```

$$\frac{1}{2}\sin(\pi\,x) - \frac{16}{9}\frac{\sin(2\,\pi\,x)}{\pi^2} - \frac{32}{225}\frac{\sin(4\,\pi\,x)}{\pi^2} - \frac{48}{1225}\frac{\sin(6\,\pi\,x)}{\pi^2}$$
$$- \frac{64}{3969}\frac{\sin(8\,\pi\,x)}{\pi^2} - \frac{80}{9801}\frac{\sin(10\,\pi\,x)}{\pi^2}$$

```
>   plot( f3(10)-x*sin(Pi*x), x=0..1 );
```

See Figure 3.15.

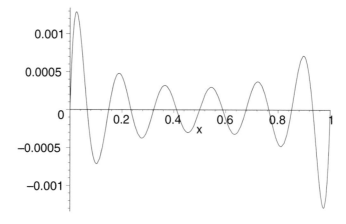

Figure 3.15: The error in 10 terms of the series for $x \sin \pi x$

```
>    int( sin(Pi*k*x)*x*sin(Pi*x), x=0..1 );
```

$$-\frac{2\,k\cos(\pi\,k) - \pi\,\sin(\pi\,k) + \pi\,k^2\sin(\pi\,k) + 2\,k}{\pi^2\,(k-1)^2\,(k+1)^2}$$

From that expression it is clear that evaluation at $k = 1$ will give problems. However, for all $k > 1$, simple evaluation (cheaper than taking limits) will work.

```
>    bad := FSS( x*sin(Pi*x)^2, x );
```

$$bad := n \rightarrow \mathrm{add}(c(k)\sin(k\,\pi\,x),\ k = 1..n)$$

```
>    bad( 10 );
```

Error, (in c) numeric exception: division by zero

```
>    good := FSS3( x*sin(Pi*x)^2, x );
```

$good := \mathbf{proc}(n)$
description
"Computes Fourier Sine Series of x*sin(Pi*x)^2"
$\cdots$
end proc

```
>    good( 10 );
```

$$\frac{4}{3}\frac{\sin(\pi\,x)}{\pi} - \frac{3}{8}\frac{\sin(2\,\pi\,x)}{\pi} - \frac{4}{15}\frac{\sin(3\,\pi\,x)}{\pi} + \frac{\frac{1}{12}\sin(4\,\pi\,x)}{\pi}$$

$$-\frac{4}{105}\frac{\sin(5\,\pi\,x)}{\pi} + \frac{\frac{1}{48}\sin(6\,\pi\,x)}{\pi} - \frac{4}{315}\frac{\sin(7\,\pi\,x)}{\pi}$$

$$+\frac{\frac{1}{120}\sin(8\,\pi\,x)}{\pi} - \frac{4}{693}\frac{\sin(9\,\pi\,x)}{\pi} + \frac{\frac{1}{240}\sin(10\,\pi\,x)}{\pi}$$

```
>    g := FSS3( Ei(x), x );
```

$g := \mathbf{proc}(n)$
description "Computes Fourier Sine Series of Ei(x)"
$\cdots$
end proc

```
>  g5 := g( 5 );
```

$$g5 := 2 \int_0^1 \sin(\pi z) \, \text{Ei}(z) \, dz \, \sin(\pi x)$$

$$+ 2 \int_0^1 \sin(2\pi z) \, \text{Ei}(z) \, dz \, \sin(2\pi x)$$

$$+ 2 \int_0^1 \sin(3\pi z) \, \text{Ei}(z) \, dz \, \sin(3\pi x)$$

$$+ 2 \int_0^1 \sin(4\pi z) \, \text{Ei}(z) \, dz \, \sin(4\pi x)$$

$$+ 2 \int_0^1 \sin(5\pi z) \, \text{Ei}(z) \, dz \, \sin(5\pi x)$$

```
>  map( evalf, g5 );
```

$$.4328262458 \sin(3.141592654\,x)$$
$$- 1.191461870 \sin(6.283185308\,x)$$
$$- .07520362192 \sin(9.424777962\,x)$$
$$- .7049139426 \sin(12.56637062\,x)$$
$$- .1096456247 \sin(15.70796327\,x)$$

```
>  w := FSS3( x^2*(1-x)*LambertW(x), x );
```

$w := \mathbf{proc}(n)$

description "Computes Fourier Sine Series of x ^2*(\
1-x)*LambertW(x)"

$\cdots$

end proc

```
>  w5 := w( 5 );
```

$$w5 := 2 \int_0^1 \sin(\pi z) \, z^2 \, (1-z) \, \text{LambertW}(z) \, dz \, \sin(\pi x)$$

$$+ 2 \int_0^1 \sin(2\pi z) \, z^2 \, (1-z) \, \text{LambertW}(z) \, dz \, \sin(2\pi x)$$

$$+ 2 \int_0^1 \sin(3\pi z) \, z^2 \, (1-z) \, \text{LambertW}(z) \, dz \, \sin(3\pi x)$$

$$+ 2 \int_0^1 \sin(4\pi z) \, z^2 \, (1-z) \, \text{LambertW}(z) \, dz \, \sin(4\pi x)$$

$$+ 2 \int_0^1 \sin(5\pi z) \, z^2 \, (1-z) \, \text{LambertW}(z) \, dz \, \sin(5\pi x)$$

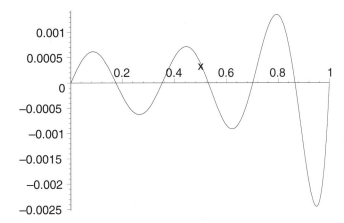

Figure 3.16: The error in 5 terms of the series for $x^2(1-x)W(x)$

```
>   w5f := evalf( w5 );
```

$$
\begin{aligned}
w5f := \; & .04899585472 \sin(3.141592654\,x) \\
& - .02898708582 \sin(6.283185308\,x) \\
& + .006104225692 \sin(9.424777962\,x) \\
& - .003229669768 \sin(12.56637062\,x) \\
& + .001449730852 \sin(15.70796327\,x)
\end{aligned}
$$

```
>   plot( w5f-x^2*(1-x)*LambertW(x), x=0..1 );
```

See Figure 3.16.

3.9.5 Solution of $y'(t) = ay(t-1)$

The following procedure uses a residue formula due to Wright [59] for the solution of the linear scalar delay equation $y'(t) = ay(t-1)$, where we assume that $a \neq -\exp(-1)$ (which is a special point), with given initial function $y(t) = f(t)$ on $0 < t \leq 1$, and where $y(0) = y_0$ is given. Wright's formula is more general than the one used here, which uses the Lambert W function [11]. These ideas are investigated theoretically in [60]. However, there are some new ideas here also, and in particular we introduce the discrete Lambert transform.

The approximate solution that is programmed is

$$
y(t) = \sum_{k=\ell}^{u} C_k e^{W_k(a)t} \, , \tag{3.7}
$$

where the C_k are chosen by Wright's residue formula to agree with the initial conditions (all sums of this form satisfy $y'(t) = a\, y(t - 1)$). If $a < 0$ is real, then the asymmetric numbering of the branches of W means that we must take $\ell = -1 - k$ to get a real answer. Since if $a > 0$ we have $W_0(a) > 0$, we know that this case is unstable; henceforth we assume $a < 0$, or at least that $\Re(a) < 0$.

The scope of the procedure in Figure 3.17 is limited to the initial functions for which the integral and residue can be calculated. The bottleneck is the integration, I suspect. As in the `Fourier_sine` procedure with which we started this book, this procedure will return an approximate answer that is a finite sum of exponential and trigonometric terms.

Programming notes

1. This procedure allows optional parameters, programmed using `Process-Options/Simple` (which itself may be improved in the next version of Maple). You may call NHFS with no initial condition at $x = 0$, and it will try to use a limit of f for that purpose. You may call NHFS without specifying which terms of the answer you require, and a default is chosen. The types and default values are set up by tables; note also the use of quoting of the option names to prevent collisions.

2. If $a > 0$, then we want to use a symmetric summation, from $\ell = -N$ to $u = N$. This is left to the user to do.

3. The program uses `try` and `catch` to control the behaviour of the program when exceptions are encountered.

4. The procedure sets `Testzero` so as to ensure that the residue is correct. It is still possible that this may fail.

5. If the input function has an integral in it, the `hasfun` test may fail.

6. The programmatic form of the final `if` statement is preferred stylistically; it is clear that something will be returned as a result.

The rest of the procedure is a straightforward implementation of the mathematics. Here is an example of its use.

```
>   restart;
>   read "D:/books/ess/programs/NHFS.mpl";
>   ans := NHFS( t, -Pi/2, exp(-t) );
```

$$ans := .009976418432\, e^{(-1.604290913\,t)} \cos(7.647192276\,t)$$
$$+ .2031279380\, e^{(-1.604290913\,t)} \sin(7.647192276\,t)$$
$$+ .6286476902 \cos(1.570796327\,t) + .2936339668 \sin(1.570796327\,t)$$

```
NHFS := proc( t::name, a::{complex(numeric),constant}, f )
   local ans, df, integralformula, k, m, n0, n1, opts,
         res, s, u, w, y0,
         Type, Default;
   description "E. M. Wright's formulas for y'(t) = a y(t-1).";
   # This method of allowing optional parameters may change with
   # the next version of Maple.  But, in the meantime, it's useful.
   Type              := table():
   Default           := table():
   Type['upper']     := integer;
   Default['upper']  := 1;
   Type['lower']     := integer;
   Default['lower']  := -2;
   Type['initial']   := {complex(numeric),constant};
   # Limit may fail, or have an error; so we try/catch it.
   try
      Default['initial'] := limit( f, t=0, right );
   catch:
      WARNING( "Default initial condition set to zero;"
               " I'm hoping that you've specified initial=y0" ) );
      Default['initial'] := 0;
   end try;
   opts := 'ProcessOptions/Simple'( 3, [args], Type, Default );
   n1 := opts['upper'];
   n0 := opts['lower'];
   y0 := opts['initial'];
   # We use the environment variables Normalizer and Testzero
   # to recognize when w*exp(w) - a is zero.
   Normalizer := simplify;
   Testzero   := b -> evalb(Normalizer(b)=0);
   # f(t) is given.
   df := unapply( diff(f,t), t );
   w := array(n0..n1):
   for k from n0 to n1 do
      w[k] := evalf(LambertW(k,a));
   end do:
   # This is H(s) from (1.8) in Wright (1947).
   integralformula := exp(s)*y0 + exp(s) * int( df(u)*exp(-s*u), u=0..1);
   if hasfun(integralformula, int) then
      error "Sorry, couldn't find the integral to start with.";
   end if;
   res := Normalizer(residue(integralformula/(s*exp(s)-a),s=LambertW(m,a)));
   userinfo(1, NHFS, "Residue is", res );
   ans := add( eval( res, LambertW(m,a)=w[k] )*exp( w[k]*t ), k=n0..n1 );
   # Use a programmatic if to make it clear we are returning something
   return   'if'( type(evalf(a),embedded_real)
                     and evalb(evalf(Re(a))<0 and n1+n0=-1),
                  evalf( evalc( Re( ans ) ) ),
                  evalf( ans ) )
end proc:
```

Figure 3.17: A Maple program to compute nonharmonic Fourier series

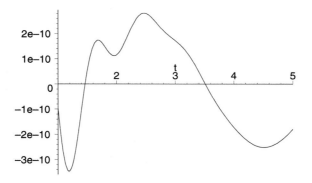

Figure 3.18: The approximate solution solves the DDE within roundoff

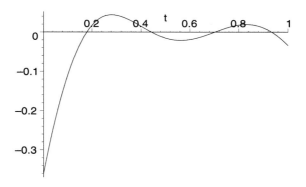

Figure 3.19: The residual in the initial function

```
>   residual := diff(ans,t) - (-Pi/2)*eval(ans,t=t-1):

>   plot( residual, t=1..5 );
```
See Figure 3.18.

By differentiating that expression and substituting it into $y'(t) + ay(t - 1)$, we see that it satisfies the differential equation to within $3 \cdot 10^{-10}$. If we work to higher precision than 10 digits, we can make the residual as small as we please.

```
>   plot( ans-exp(-t), t=0..1 );
```

See Figure 3.19. The Gibbs phenomenon [4] is clearly visible.

```
>   plot( ans, t=0..5 );
```
See Figure 3.20.

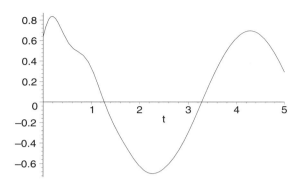

Figure 3.20: The approximate solution

> `ramp := NHFS( t, -1, t, initial=0, lower=-6, upper=5 );`

$$ramp := -.006399441292\, e^{(-3.498515212\,t)} \cos(32.88072148\,t)$$
$$+ .06014501410\, e^{(-3.498515212\,t)} \sin(32.88072148\,t)$$
$$- .009166672364\, e^{(-3.287768612\,t)} \cos(26.58047150\,t)$$
$$+ .07410937396\, e^{(-3.287768612\,t)} \sin(26.58047150\,t)$$
$$- .01437886248\, e^{(-3.020239708\,t)} \cos(20.27245764\,t)$$
$$+ .09651382300\, e^{(-3.020239708\,t)} \sin(20.27245764\,t)$$
$$- .02631875876\, e^{(-2.653191974\,t)} \cos(13.94920833\,t)$$
$$+ .1383713853\, e^{(-2.653191974\,t)} \sin(13.94920833\,t)$$
$$- .06669687276\, e^{(-2.062277730\,t)} \cos(7.588631178\,t)$$
$$+ .2454266760\, e^{(-2.062277730\,t)} \sin(7.588631178\,t)$$
$$- .3367527584\, e^{(-.3181315052\,t)} \cos(1.337235701\,t)$$
$$+ 1.415508378\, e^{(-.3181315052\,t)} \sin(1.337235701\,t)$$

> `plot( {t,ramp}, t=0..2, scaling=CONSTRAINED );`

See Figure 3.21. The poor convergence shown, the Gibbs phenomenon, can be cured by using a different approach, which I have chosen to call the discrete Lambert transform. Instead of trying to compute a finite number of the exact coefficients in the series

$$y(t) = \sum_{k=-\infty}^{\infty} c_k e^{W_k(a)t} ,$$

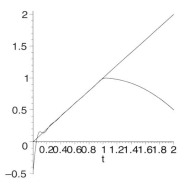

Figure 3.21: The initial ramp and the approximate solution

we instead compute a finite sum directly. That is, we compute

$$y(t) = \sum_{k=-1-N}^{N} C_k e^{W_k(a)t}$$

(remember the branch asymmetry if $a < 0$) where the C_k are chosen not as their exact counterparts c_k in the infinite series but rather so as to minimize the L_2-norm of the difference between $y(t)$ and $f(t)$ on $0 \le t \le 1$. This means that we are solving $y'(t) = ay(t-1)$ subject to $y(t) = \hat{f}(t)$ on $0 \le t \le 1$, where $\hat{f}(t)$ is supposed to be close to $f(t)$ in the L_2-norm. This gives us the problem of minimizing

$$J(C) = \int_0^1 \left| f(t) - \sum_{k=-1-N}^{N} C_k e^{W_k(a)t} \right|^2 \, dt \; .$$

We set up the minimization equations algebraically, after the methods of [46]. If we put $C_k = C_k^* + \Delta_k$, then the integrand becomes

$$\left\| \left[f(t) - \sum_k C_k^* e^{W_k(a)t} \right] - \sum_k \Delta_k e^{W_k(a)t} \right\|^2$$

which can be rewritten as

$$\left\| f(t) - \sum_k C_k^* e^{W_k(a)t} \right\|^2 - \overline{\sum_k \Delta_k e^{W_k(a)t}} \left[f(t) - \sum_k C_k^* e^{W_k(a)t} \right]$$

$$- \overline{\left[f(t) - \sum_k C_k^* e^{W_k(a)t} \right]} \sum_k \Delta_k e^{W_k(a)t} + \left\| \sum_k \Delta_k e^{W_k(a)t} \right\|^2 \; .$$

If we can find a set of C_k^* that makes

$$\int_0^1 \left[f(t) - \sum_k C_k^* e^{W_k(a)t} \right] \overline{\sum_k \Delta_k e^{W_k(a)t}} \, dt = 0 \qquad (3.8)$$

for all choices of Δ_k, then we will have shown that $J(C^* + \Delta) = J(C^*) + \int_0^1 \|\sum_k \Delta_k\|^2 \, dt$ and hence that the C_k^* provide the minimum coefficients. Certainly if this is to hold it is necessary that these equations be zero for $\Delta_k = \delta_j^k$, the Kronecker delta, for each possible j; and since that provides a linearly independent set (if $a \neq -1/e$), this is also sufficient.

This gives us $2N + 1$ linear equations in the $2N + 1$ unknowns C_j. It is now easy to see that the equations are

$$\sum_{k=-1-N}^{N} C_k \int_0^1 e^{(W_k(a) + \overline{W_j(a)})t} \, dt = \int_0^1 f(t) e^{\overline{W_j(a)}t} \, dt \text{ for } j = -1 - N, \ldots N.$$

The integrals in this matrix can be computed analytically—they are only exponentials after all—but the integrals on the right-hand side can be computed only once the function f is known, and perhaps they can be done only numerically.

Exercises

1. Show that the matrix of the last equation is Hermitian (that is, it is equal to its conjugate transpose).

2. Modify the formula (3.7) so that it is correct if $a = -\exp(-1)$, (there is a double root of $se^s = -\exp(-1)$ and thus the simple residue theory does not hold here) and then modify the procedure to deal with this case.

3. Write a procedure to find the discrete Lambert transform with respect to a of a given function $f(t)$. Re-do the example above where we used residues before, and compare the accuracy of the answer with the previous one by comparing the residual on $0 \leq t \leq 1$. You should find that the discrete Lambert transform gives a more accurate answer, although the error is more evenly spread across the interval $0 \leq t \leq 1$. Write to me when you have solved this problem. There were errors in this section in the first edition, though nobody seemed to notice but me. Let's see who's getting this far in the book, eh?

4. For which numbers a in $y'(t) = ay(t - 1)$ does the influence of perturbations in $f(t)$ die away as $t \to \infty$?

5. For which numbers a in $y'(t) = ay(t - 1)$ does the influence of persistent perturbations $y'(t) = ay(t - 1) + \varepsilon v(t)$ remain bounded? Take $\varepsilon > 0$ and $\|v\| \leq 1$.

Appendix A

A Primer on
Complex Variables

> Refutations of these misconceptions abound in the literature...
> but cannot help someone who has not read them, [or] who believes
> every elementary subject must be obvious ...
> —W. Kahan, "Mathematics Written In Sand"

Every schoolchild is taught (correctly) that there is no real number that, when squared, gives -1. Therefore, treatments of complex numbers that begin with the phrase "Let i be such that $i^2 = -1$" are fatally flawed in logic[1], and can lead to confusion, distrust, or a bankrupt pragmatism on the part of the student.

Gauss was the first to repair this fatal flaw. Other sound treatments soon followed; in particular, there is an algebraic approach based on computing in the ring of polynomials in an indeterminate (say T) modulo the polynomial $T^2 + 1$. We eschew that algebraic approach here, because it is often taken up in other computer algebra texts. Instead, we follow Gauss. We can paraphrase his work as follows. Consider the algebra of *pairs of real numbers* given by the transformation rules

$$(x_1, y_1) + (x_2, y_2) = (x_1 + x_2, y_1 + y_2) \tag{A.1}$$

$$(x_1, y_1) \cdot (x_2, y_2) = (x_1 x_2 - y_1 y_2, x_1 y_2 + y_1 x_2). \tag{A.2}$$

That last rule, which we can check defines a commutative multiplication, is more easily understood in *polar coordinates*, which we will look at after we have defined the two-argument arctan function in a later section. These algebraic rules

---

[1] Consider the equally flawed beginning "Let N be the largest integer" of Perron's paradox. From there, one can deduce that $N = 1$, or other nonsense.

have the following consequences:

$$(x_1, 0) + (x_2, 0) = (x_1 + x_2, 0)$$
$$(x_1, 0) \cdot (x_2, 0) = (x_1 x_2, 0) .$$

That is, this set of number pairs contains an isomorphic copy of the ring of real numbers (later we will see that the division operation, making it a field, is also copied). By overloading notation, we write 1 for $(1, 0)$, and indeed x for $(x, 0)$. We also overload the arithmetic operators so that we may add and multiply real and complex numbers in the natural fashion: $x \cdot (x_1, y_1)$ we take to mean $(x, 0) \cdot (x_1, y_1)$, etc. Defining subtraction is easy: $(x_1, y_1) - (x_2, y_2) = (x_1, y_1) + (-x_2, -y_2)$. Defining division by $z \neq 0$ can be done by noticing that

$$(x_1, y_1) \cdot \left(\frac{x_1}{x_1^2 + y_1^2}, \frac{-y_1}{x_1^2 + y_1^2} \right) = (1, 0) = 1 , \qquad (A.3)$$

i.e. that we can easily compute z^{-1} given $z \neq 0$; we then define a/b to be ab^{-1}. We call the set of number pairs with this algebra *the complex numbers*, and denote it by $\mathbb{C}$.

It is possible to notice that

$$(0, 1) \cdot (0, 1) = (-1, 0) \qquad (A.4)$$

which is the isomorphic copy of the real number -1, and so we may *define*

$$i := (0, 1) . \qquad (A.5)$$

Therefore, in this algebra of pairs of real numbers, there is a number pair that, when squared, gives us the isomorphic copy of -1. By the overloaded notation introduced earlier, we may write (x, y) as $(x, 0) + (0, y)$ or $x \cdot (1, 0) + (0, 1) \cdot y$ or $x + iy$. We could equally well have defined i to be $(0, -1)$, which is also equal to -1 when squared, but by convention we call this number $-i$.

Maple's representation of a complex number $a + bi$ is as the function call `Complex(a,b)`, though it prints as `a + bI`.

A.1 Polar Coordinates and the Two-Argument Arctan Function

The rule (A.2) for multiplication of complex numbers is more easily understood in polar coordinates. Consider the complex number $z = (x, y)$ to define the Cartesian coordinates of a point in the plane. The polar coordinates $[\rho, \theta]$ of that point are given by

$$\rho = \sqrt{x^2 + y^2} \qquad (A.6)$$

and

$$\theta = \arctan(x, y) . \tag{A.7}$$

You may have seen that before, perhaps expressed as $\tan^{-1}(y/x)$, and have been expected to work out for yourself which quadrant the angle was in and therefore get the angle correct in $-\pi < \theta \le \pi$. This is because, of course, $y/x = (-y)/(-x)$; and hence points in quadrants I and III, II and IV are indistinguishable once this ratio has been taken.

The purpose of the two-argument arctan function (just `arctan(x,y)` in Maple, `atan2(x,y)` in MATLAB and FORTRAN) is to save you the trouble. The two-argument arctan takes the Cartesian coordinates of a point in the plane and returns the polar angle of the point, in the interval $(-\pi, \pi]$.

Remark. By convention, the angle is taken in $(-\pi, \pi]$. This defines the *branch cut* and its *closure* (by which we mean that the angle is continuous or "closed" as you approach the negative real axis from above, or in a counterclockwise direction around the origin) for the angle or "argument" function. This is only a convention, and could have been chosen anywhere else (say $[0, 2\pi)$)—but it wasn't. This convention, called "counter-clockwise closure" or "CCC" by [36] is by now nearly universal in computer languages.

In polar coordinates, the multiplication rule (A.2) becomes

$$[\rho_1, \theta_1] \cdot [\rho_2, \theta_2] = [\rho_1 \rho_2, \theta_1 + \theta_2] \tag{A.8}$$

as is easily seen by converting (x_1, y_1) and (x_2, y_2) to polar coordinates and using the trig identities

$$\cos(\theta_1 + \theta_2) = \cos \theta_1 \cos \theta_2 - \sin \theta_1 \sin \theta_2$$

$$\sin(\theta_1 + \theta_2) = \cos \theta_1 \sin \theta_2 + \cos \theta_2 \sin \theta_1 .$$

Here we do not care if the angle sum $\theta_1 + \theta_2$ is outside the range $(-\pi, \pi]$, because in translating back to Cartesian coordinates via

$$x = \rho \cos \theta \tag{A.9}$$

$$y = \rho \sin \theta \tag{A.10}$$

all that matters is θ modulo 2π. Sadly, in polar coordinates, addition is now more complicated ... so we need both systems.

A.2   The Exponential Function

We are now ready to define the exponential function of a complex variable. We write

$$e^z = \exp(z) := \sum_{k=0}^{\infty} \frac{z^k}{k!} . \tag{A.11}$$

In standard textbooks, for example the beautiful [29], we find proofs that this series converges for all finite z and that this is therefore well-defined. From this definition, with some work, we may deduce Euler's formula:

$$e^{x+iy} = e^x(\cos y + i\sin y)\,,\tag{A.12}$$

from which many properties of the exponential function can be shown, including the relation between the Cartesian and polar coordinates:

$$z = x + iy = \rho\cos\theta + i\rho\sin\theta = \rho e^{i\theta}\,.\tag{A.13}$$

The property that concerns us most here is that

$$e^{2\pi ik} = \cos(2\pi k) + i\sin(2\pi k) = 1\tag{A.14}$$

for any integer k: that is, the exponential function is many-to-one. Therefore, its *functional inverse* is multi-valued.

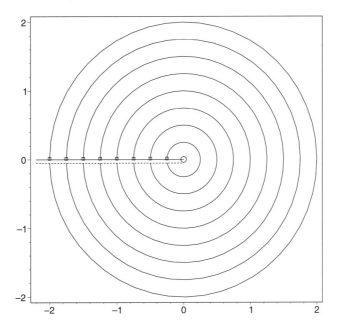

Figure A.1: Circular arcs in the complex domain of logarithm

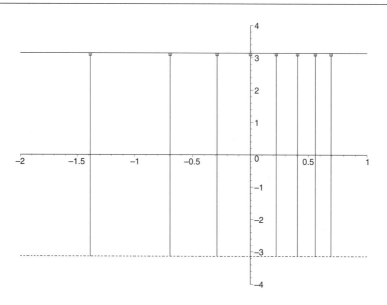

Figure A.2: The principal complex range of logarithm

A.3   The Natural Logarithm

We define the *principal branch* of the natural logarithm of $z = x + iy = \rho e^{i\theta}$ to be (from the polar coordinate form)

$$\ln z = \ln \rho e^{i\theta} := \ln \rho + i\theta = \ln \rho + i \arctan(x, y) . \qquad \text{(A.15)}$$

This therefore inherits the same branch cut as the two-argument arctan function, namely along the negative real axis, with closure from above. See Figures A.1 and A.2. I have indicated closure on the top of the negative real axis in the domain of the logarithm by marking it as a solid line, and putting a dashed line underneath to indicate the domain is open there, and putting boxes touching the top of the axis on the ends of the lines through the domain. The corresponding lines in the principal range of the logarithm are not circular arcs but rather straight line segments. Closure is again indicated by a mixture of solid lines, boxes, and dashed lines. For other visual descriptions of branch cut closures, see [54] or its electronic edition. For more discussion of these closures, see [13].

Once we have defined a logarithm, we may define general powers as follows.

$$z^a := e^{a \ln z} . \qquad \text{(A.16)}$$

Because of equation (A.14), we could equally well have chosen

$$\ln_k z = \ln z + 2\pi i k \qquad (A.17)$$

for any other integer k to be our canonical logarithm. We do not. Following the de facto standard, we choose $k = 0$, and thus $-\pi < \theta \le \pi$, to be the one. Every computer algebra language and numerical language follows this standard and takes the complex logarithm to have its imaginary part (also called the "argument" of z) in this range.

With this definition, $(-8)^{1/3} = 1 + i\sqrt{3}$, and not -2. If you wish *real-valued* cube roots (or other rational roots), use `surd`.

```
>   (-8)^(1/3);
```
$$(-8)^{(1/3)}$$

```
>   simplify( % );
```
$$1 + I\sqrt{3}$$

```
>   surd( -8, 3 );
```
$$-2$$

'And yet this thing is, to my confined and but part-conceiving intellects, a surd: an irrational incogitable.'
—E. R. Eddison, *The Mezentian Gate*, Chapter 29.

Remark. Several rules that we learned in high school that are valid for positive reals are not necessarily valid for complex numbers. In particular, it is not true that $\ln(ab) = \ln a + \ln b$, or that $\ln z^2 = 2 \ln z$, or that $(z^a)^b = z^{ab}$. Corrections to these identities are presented in Table A.1.

Exercises

1. Show that $\exp(\ln z) = z$ for all complex z (so *some* high-school identities are true).

2. Show that there exists complex (in fact, real) z such that $\sqrt{1/z} \ne 1/\sqrt{z}$.

3. Show that there exists complex (in fact, real) z, w such that $\sqrt{zw} \ne \sqrt{z} \cdot \sqrt{w}$.

4. Show that, with the Maple definitions,

$$\arcsin z = \arctan \frac{z}{\sqrt{1-z^2}} + \pi\mathcal{K}(-\ln(1+z)) - \pi\mathcal{K}(-\ln(1-z)). \quad (A.18)$$

$$\mathcal{K}(z) = \left\lceil \frac{\Im(z) - \pi}{2\pi} \right\rceil$$

$$\ln e^z = z - 2\pi i \mathcal{K}(z)$$

$$\ln(z_1 z_2) = \ln z_1 + \ln z_2 - 2\pi i \mathcal{K}(\ln z_1 + \ln z_2)$$

$$\ln(z^a) = a \ln z + 2\pi i \mathcal{K}(a \ln z)$$

$$\mathcal{K}(n \ln z) = 0 \; \forall z \text{ iff} -1 < n \leq 1$$

$$(z^a)^b = z^{ab} e^{2\pi i b \mathcal{K}(a \ln z)}$$

$$\sqrt{z^2} = z \operatorname{csgn}(z) := z e^{\pi i \mathcal{K}(2 \ln z)}$$

$$\left(z^n\right)^{1/n} = z \, C_n(z) := z e^{2\pi i \mathcal{K}(n \ln z)/n}$$

Table A.1: Some identities for the complex logarithm

A.4   Trig Functions and Hyperbolic Functions

Once we have the exponential function, it is possible to define the complex trig functions:

$$\sin z = \frac{e^{iz} - e^{-iz}}{2i}$$

$$\cos z = \frac{e^{iz} + e^{-iz}}{2}$$

and from thence all the other trig functions: $\tan z := \sin z / \cos z$, $\csc z := 1 / \sin z$, $\sec z := 1 / \cos z$, and $\cot z := 1 / \tan z$.

A.5   Inverse Trigs and Hyperbolics

By solving $y = \sin z$ for z, we arrive at an expression for $\arcsin y$ that depends on logarithms. The following so-called principal expressions are carefully chosen to give good branch cuts and closures on the branch cuts so that we have agreement with many numerical languages.

```
>   restart;
>   convert( arcsin(z), ln );
```
$$-I \ln(\sqrt{1 - z^2} + I z)$$
```
>   convert( arccos(z), ln );
```
$$-I \ln(z + I \sqrt{1 - z^2})$$

```
> convert( arctan(z), ln );
```
$$\frac{1}{2} I \left(\ln(1 - I\,z) - \ln(1 + I\,z) \right)$$

```
> convert( arccsc(z), ln );
```
$$-I \ln\left(\frac{I}{z} + \sqrt{1 - \frac{1}{z^2}}\right)$$

```
> convert( arcsec(z), ln );
```
$$-I \ln\left(\frac{1}{z} + I\sqrt{1 - \frac{1}{z^2}}\right)$$

```
> convert( arccot(z), ln );
```
$$\frac{1}{2}\pi - \frac{1}{2} I \left(\ln(1 - I\,z) - \ln(1 + I\,z) \right)$$

```
> convert( arcsinh(z), ln );
```
$$\ln(z + \sqrt{z^2 + 1})$$

```
> convert( arccosh(z), ln );
```
$$\ln(z + \sqrt{z - 1}\,\sqrt{z + 1})$$

```
> convert( arctanh(z), ln );
```
$$\frac{1}{2}\ln(z + 1) - \frac{1}{2}\ln(1 - z)$$

```
> convert( arccsch(z), ln );
```
$$\ln\left(\frac{1}{z} + \sqrt{1 + \frac{1}{z^2}}\right)$$

```
> convert( arcsech(z), ln );
```
$$\ln\left(\frac{1}{z} + \sqrt{\frac{1}{z} - 1}\,\sqrt{\frac{1}{z} + 1}\right)$$

```
> convert( arccoth(z), ln );
```
$$\frac{1}{2}\ln(z + 1) - \frac{1}{2}\ln(z - 1)$$

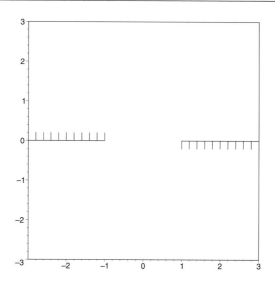

Figure A.3: The domain of arcsin, arccos, and arctan

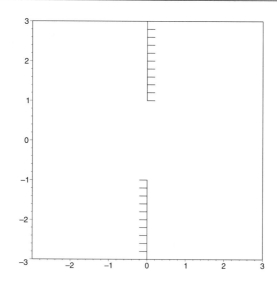

Figure A.4: The domain of arctan, arccot, and arcsinh

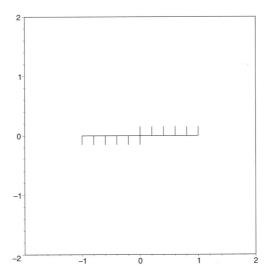

Figure A.5: The domain of arccsc and arcsec

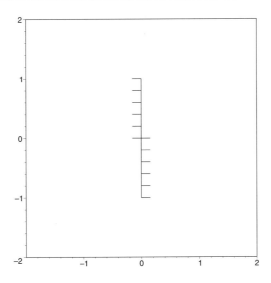

Figure A.6: The domain of arccsch

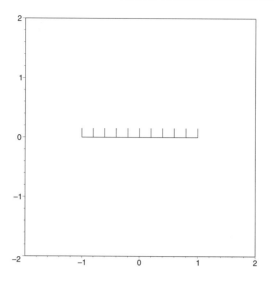

Figure A.7: The domain of arccoth

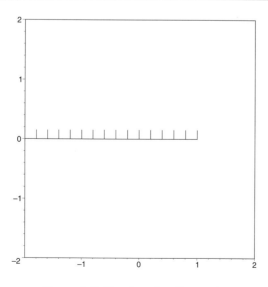

Figure A.8: The domain of arccosh

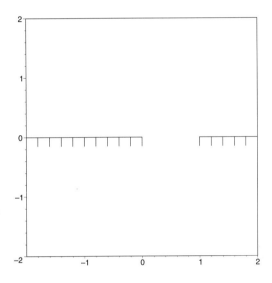

Figure A.9: The domain of arcsech

Bibliography

[1] E. J. Barbeau. *Polynomials*. Problem Books in Mathematics. Springer-Verlag, 1989.

[2] C. M. Bender and S. A. Orszag. *Advanced Mathematical Methods for Scientists and Engineers*. McGraw-Hill, 1978.

[3] Wolf-Jürgen Beyn. *Numerical Methods for Dynamical Systems*, pages 175–236. Oxford Science Publications, 1991.

[4] William E. Boyce and Richard C. DiPrima. *Elementary Differential Equations and Boundary Value Problems*. Wiley, 2nd edition, 1969.

[5] Michael W. Chamberlain. Heart to bell. *College Mathematics Journal*, 25(1):34, January 1994.

[6] Bruce W. Char, Keith O. Geddes, Gaston H. Gonnet, Benton L. Leong, Michael B. Monagan, and Stephen M. Watt. *The Maple V Language Reference Manual*. Springer-Verlag, 1991.

[7] Robert M. Corless. Continued fractions and chaos. *American Mathematical Monthly*, 99(3):203–215, 1992.

[8] Robert M. Corless. Error backward. In Peter Kloeden and Ken Palmer, editors, *Contemporary Mathematics*, volume 172, pages 31–62. American Mathematical Society, 1994.

[9] Robert M. Corless. A review of *Modern Computer Algebra*. SIGSAM *Bulletin*, 34(135):8–13, 2001.

[10] Robert M. Corless, James H. Davenport, David J. Jeffrey, and Stephen M. Watt. According to Abramowitz & Stegun, or Arcoth needn't be uncouth. SIGSAM *Bulletin*, 34(2, issue 132):58–65, June 2000.

[11] Robert M. Corless, Gaston H. Gonnet, D. E. G. Hare, David J. Jeffrey, and Donald E. Knuth. On the Lambert W function. *Advances in Computational Mathematics*, 5:329–359, 1996.

[12] Robert M. Corless and Michael C. Haslam. More ghost curves of Chebyshev polynomials. Technical report, ORCCA, in progress, 2001.

[13] Robert M. Corless, James H. Davenport, David J. Jeffrey, and Stephen M. Watt. According to Abramowitz and Stegun, or, Arccoth needn't be uncouth. SIGSAM BULLETIN: *Communications on Computer Algebra*, 34(2), June 2000. Ontario Research Centre for Computer Algebra Technical Report TR-00-17, at http://www.orcca.on.ca/TechReports.

[14] Robert M. Corless and David J. Jeffrey. Well, it isn't quite that simple.... *SIGSAM Bulletin*, 26(3):2–6, August 1992.

[15] Robert M. Corless and David J. Jeffrey. The Turing factorization of a rectangular matrix. SIGSAM *Bulletin (Communications in Computer Algebra)*, 31(3):20–28, September 1997.

[16] Robert M. Corless and David J. Jeffrey. Elementary Riemann surfaces. SIGSAM *Bulletin*, 32(1):11–17, March 1998.

[17] Robert M. Corless and David J. Jeffrey. On the Lambert W function and the Wright ω function. Technical Report TR-00-12, at http://www.orcca.on.ca/TechReports, Ontario Research Centre for Computer Algebra, 2000.

[18] Robert M. Corless, David J. Jeffrey, Michael B. Monagan, and Pratibha. Two perturbation calculations in fluid mechanics using large-expression management. *J. Symbolic Computation*, 23:427–443, 1997.

[19] David Cox, John Little, and Donal O'Shea. *Ideals, Varieties, and Algorithms.* Springer-Verlag, 1992.

[20] Temple H. Fay. The butterfly curve. *American Mathematical Monthly*, 96(5):442–443, May 1989.

[21] Walter Gander and Jiří Hřebíček. *Solving Problems in Scientific Computing Using Maple and MATLAB.* Springer, 1993.

[22] Keith O. Geddes, Stephen R. Czapor, and George Labahn. *Algorithms for Computer Algebra.* Kluwer, 1992.

[23] David Goldberg. What every computer scientist should know about floating-point arithmetic. *ACM Computing Surveys*, 23(1):5–48, 1991.

[24] Gene Golub and Charles Van Loan. *Matrix Computations.* Johns Hopkins, 2nd edition, 1989.

[25] Ronald L. Graham, Donald E. Knuth, and Oren Patashnik. *Concrete Mathematics.* Addison-Wesley, 1994.

[26] Ernst Hairer, Syvert P. Nørsett, and Gerhard Wanner. *Solving Ordinary Differential Equations*, volume I of *Computational Mathematics*. Springer-Verlag, 1987.

[27] Godfrey Harold Hardy. *Divergent Series.* Clarendon Press, Oxford, 1949.

[28] André Heck. *Introduction to Maple.* Springer-Verlag, 1993.

[29] Peter Henrici. *Applied and Computational Complex Analysis*, volume I. Wiley-Interscience, 1977.

[30] Desmond J. Higham and Nicholas J. Higham. *The MATLAB guide.* SIAM, 2000.

[31] Nicholas J. Higham. *Accuracy and Stability of Numerical Algorithms*. Society for Industrial and Applied Mathematics, Philadelphia, PA, USA, 1996.

[32] John H. Hubbard and Beverly H. West. *Differential Equations: a Dynamical Systems Approach*. Springer-Verlag, 1991.

[33] T. E. Hull, W. H. Enright, B. M. Fellen, and A. E. Sedgwick. Comparing numerical methods for ordinary differential equations. *SIAM J. Numer. Anal.*, 9:603–637, 1972.

[34] David J. Jeffrey. The importance of being continuous. *Mathematics Magazine*, 67(4):294–300, 1994.

[35] William M. Kahan. Handheld calculator evaluates integrals. *Hewlett-Packard Journal*, pages 23–32, August 1980.

[36] William M. Kahan. Branch cuts for complex elementary functions, or, much ado about nothing's sign bit. In M. J. D. Powell and A. Iserles, editors, *The state of the art in numerical analysis: Proceedings of the Joint IMA/SIAM Conference*. Oxford University Press, April 1986.

[37] William M. Kahan. Lecture notes on the status of IEEE754. *http://www.cs.berkeley.edu/~wkahan/IEEE754status/IEEE754.ps*, May 1996.

[38] R. A. Knoebel. Exponentials reiterated. *American Mathematical Monthly*, 88:235–252, 1981.

[39] Donald E. Knuth and B. Pittel. "A recurrence related to trees". *Proc. Amer. Math. Soc.*, 105:335–349, 1989.

[40] T. W. Körner. *Fourier Analysis*. Cambridge University Press, 1989.

[41] T. Y. Li, Tim Sauer, and J. A. Yorke. The cheater's homotopy. *SIAM Journal of Numerical Analysis*, 26(5):1241–1251, October 1989.

[42] John McCleary. How not to prove Fermat's last theorem. *Amer. Math. Monthly*, 96(5):410–420, May 1989.

[43] Christian Mittermaier, Wolfgang Schreiner, and Franz Winkler. A parallel symbolic-numerical approach to algebraic curve plotting. In *Proceedings of CASC*, pages 301–314. Springer, 2000.

[44] M. B. Monagan, K. O. Geddes, K. M. Heal, G. Labahn, S. M. Vorkoetter, J. McCarron, and P. DeMarco. *Maple 7 Programming Guide*. Waterloo Maple, Inc., 2001.

[45] Alexander Morgan. *Solving Polynomial Systems Using Continuation for Engineering and Scientific Problems*. Prentice-Hall, 1987.

[46] Ivan Niven. *Maxima and Minima without Calculus*, volume 6 of *Dolciani Mathematical Expositions*. Mathematical Association of America, 1981.

[47] American National Standards Institute/Institute of Electrical and Electronic Engineers. *IEEE Standard for Binary Floating-Point Arithmetic*. ANSI/IEEE Std 754-1985, 1985.

[48] Theodore J. Rivlin. *Chebyshev Polynomials*. Wiley-Interscience, 1990.

[49] Jr. Robert E. O'Malley. *Singular Perturbation Methods for Ordinary Differential Equations*, volume 89 of *Applied Mathematical Sciences*. Springer-Verlag, 1991.

[50] Bruno Salvy. Efficient programming in Maple: A case study. *SIGSAM Bulletin*, 27(2):1–12, April 1993.

[51] J. Rafael Sendra and Franz Winkler. Symbolic parameterization of curves. *Journal of Symbolic Computation*, 12(6):607–631, 1991.

[52] Lawrence F. Shampine and Robert M. Corless. Initial value problems for ODEs in problem solving environments. *J. Computational & Applied Mathematics*, 125(1–2):31–40, 2000.

[53] Helen Skala. Contour maps—a visual experience. *The College Mathematics Journal*, 22(3):241–244, May 1991.

[54] Guy L. Steele. *Common Lisp the Language*. Digital Press, 1990.

[55] Joachim von zur Gathen and Jürgen Gerhard. *Modern Computer Algebra*. Cambridge University Press, 1999.

[56] Volker Weispfenning. Comprehensive Gröbner bases. *Journal of Symbolic Computation*, 14:1–29, July 1992.

[57] Ernst Joachim Weniger. *Nonlinear Sequence Transformations for the acceleration of convergence and the summation of divergent series*, volume 10 of *Computer Physics Reports*. North-Holland, December 1989.

[58] James H. Wilkinson. *The Perfidious Polynomial*, pages 1–28. Mathematical Association of America, 1984.

[59] Edward Maitland Wright. The linear difference-differential equation with constant coefficients. *Proc. Royal Soc. Edinburgh A*, LXII:387–393, 1947.

[60] Hualiang Zhang. *Non-harmonic Fourier series and applications*. PhD thesis, University of Western Ontario, London, CANADA, January 2000.

Index